CORRIGENDUM
ENVIRONMENTAL HEALTH CRITERIA NO. 231
BENTONITE, KAOLIN AND SELECTED
CLAY MINERALS

Add after the second paragraph on page xiv:

The draft EHC on Bentonite, Kaolin and Selected Clay Minerals was sent for review to IPCS National Contact Points and Participating Institutions, as well as to identified experts. Comments were received from:

M. Baril, Institut de Recherché en Santé et en Sécurité du Travail, Canada
R. Benson, US Environmental Protection Agency, USA
R. Chhabra, National Institute of Environmental Health Sciences, USA
E. Frantik, National Institute of Public Health, Czech Republic
H. Gibb, Sciences International, Alexandria, VA, USA
R. F. Hertel, Bundesinstitut für Risikobewertung, Germany
F. Javier Huertas, CSIC Estacion Experimental del Zaidin, Granada, Spain
G. Nordberg, Umeå University, Umeå, Sweden
P. Isaac Rabbani, Food & Drug Administration, USA
H. Savolainen, Ministry of Social Affairs & Health, Finland
P. Schulte, National Institute of Occupational Safety and Health, USA
J.L. Stauber, CSIRO Energy Technology, Australia
A. Tossavainen, Institute of Occupational Health, Finland
B. Velde, Laboratoire de géologie de l'école normale supérieure de Paris, France
R.B. Williams, formerly US Environmental Protection Agency, and Regional Office for the Americas of the World Health Organization

Environmental Health Criteria 231

BENTONITE, KAOLIN, AND SELECTED CLAY MINERALS

First draft prepared by Dr Zoltán Adamis, József Fodor National Center for Public Health, National Institute of Chemical Safety, Budapest, Hungary; and Dr Richard B. Williams, US Environmental Protection Agency, Washington, DC, and Regional Office for the Americas of the World Health Organization

Published under the joint sponsorship of the United Nations Environment Programme, the International Labour Organization, and the World Health Organization, and produced within the framework of the Inter-Organization Programme for the Sound Management of Chemicals.

World Health Organization
Geneva, 2005

The **International Programme on Chemical Safety (IPCS)**, established in 1980, is a joint venture of the United Nations Environment Programme (UNEP), the International Labour Organization (ILO), and the World Health Organization (WHO). The overall objectives of the IPCS are to establish the scientific basis for assessment of the risk to human health and the environment from exposure to chemicals, through international peer review processes, as a prerequisite for the promotion of chemical safety, and to provide technical assistance in strengthening national capacities for the sound management of chemicals.

The **Inter-Organization Programme for the Sound Management of Chemicals (IOMC)** was established in 1995 by UNEP, ILO, the Food and Agriculture Organization of the United Nations, WHO, the United Nations Industrial Development Organization, the United Nations Institute for Training and Research, and the Organisation for Economic Co-operation and Development (Participating Organizations), following recommendations made by the 1992 UN Conference on Environment and Development to strengthen cooperation and increase coordination in the field of chemical safety. The purpose of the IOMC is to promote coordination of the policies and activities pursued by the Participating Organizations, jointly or separately, to achieve the sound management of chemicals in relation to human health and the environment.

WHO Library Cataloguing-in-Publication Data

Bentonite, kaolin, and selected clay minerals.

(Environmental health criteria ; 231)

1.Bentonite - toxicity 2.Kaolin - toxicity 3.Aluminum silicates - toxicity 4.Environmental exposure 5.Risk assessment I.International Programme for Chemical Safety II.Series

ISBN 92 4 157231 0   (LC/NLM classification: QV 65)
ISSN 0250-863X

©World Health Organization 2005

This document was technically and linguistically edited by Marla Sheffer, Ottawa, Canada, and printed by Wissenchaftliche Verlagsgesellschaft mbH, Stuttgart, Germany.

CONTENTS

ENVIRONMENTAL HEALTH CRITERIA FOR
BENTONITE, KAOLIN, AND SELECTED CLAY MINERALS

NOTE TO READERS OF THE CRITERIA MONOGRAPHS

Every effort has been made to present information in the criteria monographs as accurately as possible without unduly delaying their publication. In the interest of all users of the Environmental Health Criteria monographs, readers are requested to communicate any errors that may have occurred to the Director of the International Programme on Chemical Safety, World Health Organization, Geneva, Switzerland, in order that they may be included in corrigenda.

Environmental Health Criteria

PREAMBLE

Objectives

In 1973 the WHO Environmental Health Criteria Programme was initiated with the following objectives:

(i)   to assess information on the relationship between exposure to environmental pollutants and human health, and to provide guidelines for setting exposure limits;
(ii)   to identify new or potential pollutants;
(iii)   to identify gaps in knowledge concerning the health effects of pollutants;
(iv)   to promote the harmonization of toxicological and epidemiological methods in order to have internationally comparable results.

The first Environmental Health Criteria (EHC) monograph, on mercury, was published in 1976, and since that time an ever-increasing number of assessments of chemicals and of physical effects have been produced. In addition, many EHC monographs have been devoted to evaluating toxicological methodology, e.g., for genetic, neurotoxic, teratogenic and nephrotoxic effects. Other publications have been concerned with epidemiological guidelines, evaluation of short-term tests for carcinogens, biomarkers, effects on the elderly and so forth.

Since its inauguration the EHC Programme has widened its scope, and the importance of environmental effects, in addition to health effects, has been increasingly emphasized in the total evaluation of chemicals.

The original impetus for the Programme came from World Health Assembly resolutions and the recommendations of the 1972 UN Conference on the Human Environment. Subsequently the work became an integral part of the International Programme on Chemical Safety (IPCS), a cooperative programme of UNEP, ILO and WHO. In this manner, with the strong support of the new partners, the importance of occupational health and environmental effects was fully recognized. The EHC monographs have become widely established, used and recognized throughout the world.

The recommendations of the 1992 UN Conference on Environment and Development and the subsequent establishment of the Intergovernmental Forum on Chemical Safety with the priorities for action in the six programme areas of Chapter 19, Agenda 21, all lend further weight to the need for EHC assessments of the risks of chemicals.

Scope

The criteria monographs are intended to provide critical reviews on the effect on human health and the environment of chemicals and of combinations of chemicals and physical and biological agents. As such, they include and review studies that are of direct relevance for the evaluation. However, they do not describe *every* study carried out. Worldwide data are used and are quoted from original studies, not from abstracts or reviews. Both published and unpublished reports are considered, and it is incumbent on the authors to assess all the articles cited in the references. Preference is always given to published data. Unpublished data are used only when relevant published data are absent or when they are pivotal to the risk assessment. A detailed policy statement is available that describes the procedures used for unpublished proprietary data so that this information can be used in the evaluation without compromising its confidential nature (WHO (1990) Revised Guidelines for the Preparation of Environmental Health Criteria Monographs. PCS/90.69, Geneva, World Health Organization).

In the evaluation of human health risks, sound human data, whenever available, are preferred to animal data. Animal and *in vitro* studies provide support and are used mainly to supply evidence missing from human studies. It is mandatory that research on human subjects is conducted in full accord with ethical principles, including the provisions of the Helsinki Declaration.

The EHC monographs are intended to assist national and international authorities in making risk assessments and subsequent risk management decisions. They represent a thorough evaluation of risks and are not, in any sense, recommendations for regulation or standard setting. These latter are the exclusive purview of national and regional governments.

Content

The layout of EHC monographs for chemicals is outlined below.

- Summary — a review of the salient facts and the risk evaluation of the chemical
- Identity — physical and chemical properties, analytical methods
- Sources of exposure
- Environmental transport, distribution and transformation
- Environmental levels and human exposure
- Kinetics and metabolism in laboratory animals and humans
- Effects on laboratory mammals and *in vitro* test systems
- Effects on humans
- Effects on other organisms in the laboratory and field
- Evaluation of human health risks and effects on the environment
- Conclusions and recommendations for protection of human health and the environment
- Further research
- Previous evaluations by international bodies, e.g., IARC, JECFA, JMPR

Selection of chemicals

Since the inception of the EHC Programme, the IPCS has organized meetings of scientists to establish lists of priority chemicals for subsequent evaluation. Such meetings have been held in Ispra, Italy, 1980; Oxford, United Kingdom, 1984; Berlin, Germany, 1987; and North Carolina, USA, 1995. The selection of chemicals has been based on the following criteria: the existence of scientific evidence that the substance presents a hazard to human health and/or the environment; the possible use, persistence, accumulation or degradation of the substance shows that there may be significant human or environmental exposure; the size and nature of populations at risk (both human and other species) and risks for environment; international concern, i.e., the substance is of major interest to several countries; adequate data on the hazards are available.

If an EHC monograph is proposed for a chemical not on the priority list, the IPCS Secretariat consults with the Cooperating Organizations and all the Participating Institutions before embarking on the preparation of the monograph.

Procedures

The order of procedures that result in the publication of an EHC monograph is shown in the flow chart on p. xii. A designated staff member of IPCS, responsible for the scientific quality of the document, serves as Responsible Officer (RO). The IPCS Editor is responsible for layout and language. The first draft, prepared by consultants or, more usually, staff from an IPCS Participating Institution, is based on extensive literature searches from reference databases such as Medline and Toxline.

The draft document, when received by the RO, may require an initial review by a small panel of experts to determine its scientific quality and objectivity. Once the RO finds the document acceptable as a first draft, it is distributed, in its unedited form, to well over 150 EHC contact points throughout the world who are asked to comment on its completeness and accuracy and, where necessary, provide additional material. The contact points, usually designated by governments, may be Participating Institutions, IPCS Focal Points or individual scientists known for their particular expertise. Generally some four months are allowed before the comments are considered by the RO and author(s). A second draft incorporating comments received and approved by the Director, IPCS, is then distributed to Task Group members, who carry out the peer review, at least six weeks before their meeting.

The Task Group members serve as individual scientists, not as representatives of any organization, government or industry. Their function is to evaluate the accuracy, significance and relevance of the information in the document and to assess the health and environmental risks from exposure to the chemical. A summary and recommendations for further research and improved safety aspects are also required. The composition of the Task Group is dictated by the range of expertise required for the subject of the meeting and by the need for a balanced geographical distribution.

EHC PREPARATION FLOW CHART

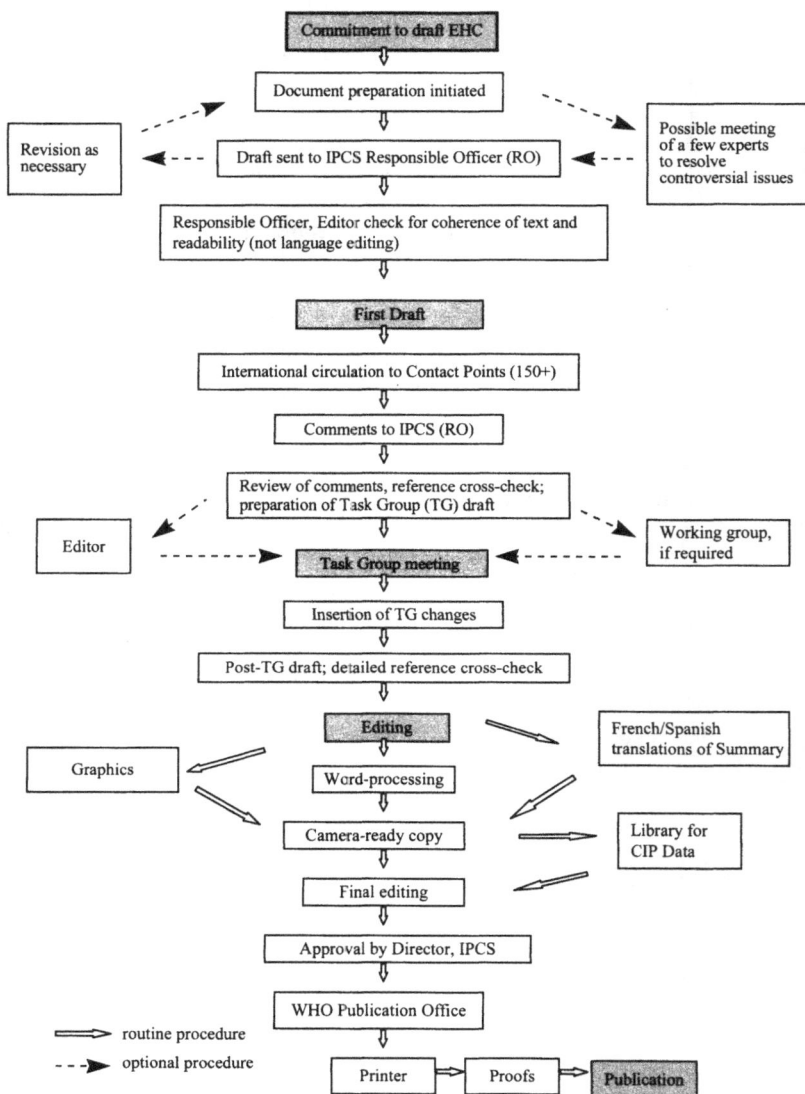

Commitment to draft EHC
⇩
Document preparation initiated
⇩
Draft sent to IPCS Responsible Officer (RO)
⇩
Responsible Officer, Editor check for coherence of text and readability (not language editing)
⇩
First Draft
⇩
International circulation to Contact Points (150+)
⇩
Comments to IPCS (RO)
⇩
Review of comments, reference cross-check; preparation of Task Group (TG) draft
⇩
Task Group meeting
⇩
Insertion of TG changes
⇩
Post-TG draft; detailed reference cross-check
⇩
Editing
⇩
Word-processing
⇩
Camera-ready copy
⇩
Final editing
⇩
Approval by Director, IPCS
⇩
WHO Publication Office
⇩
Printer → Proofs → Publication

Revision as necessary

Possible meeting of a few experts to resolve controversial issues

Editor

Working group, if required

Graphics

French/Spanish translations of Summary

Library for CIP Data

⟹ routine procedure
- - -▶ optional procedure

The three cooperating organizations of the IPCS recognize the important role played by nongovernmental organizations. Representatives from relevant national and international associations may be invited to join the Task Group as observers. Although observers may provide a valuable contribution to the process, they can speak only at the invitation of the Chairperson. Observers do not participate in the final evaluation of the chemical; this is the sole responsibility of the Task Group members. When the Task Group considers it to be appropriate, it may meet *in camera*.

All individuals who as authors, consultants or advisers participate in the preparation of the EHC monograph must, in addition to serving in their personal capacity as scientists, inform the RO if at any time a conflict of interest, whether actual or potential, could be perceived in their work. They are required to sign a conflict of interest statement. Such a procedure ensures the transparency and probity of the process.

When the Task Group has completed its review and the RO is satisfied as to the scientific correctness and completeness of the document, it then goes for language editing, reference checking and preparation of camera-ready copy. After approval by the Director, IPCS, the monograph is submitted to the WHO Office of Publications for printing. At this time a copy of the final draft is sent to the Chairperson and Rapporteur of the Task Group to check for any errors.

It is accepted that the following criteria should initiate the updating of an EHC monograph: new data are available that would substantially change the evaluation; there is public concern for health or environmental effects of the agent because of greater exposure; an appreciable time period has elapsed since the last evaluation.

All Participating Institutions are informed, through the EHC progress report, of the authors and institutions proposed for the drafting of the documents. A comprehensive file of all comments received on drafts of each EHC monograph is maintained and is available on request. The Chairpersons of Task Groups are briefed before each meeting on their role and responsibility in ensuring that these rules are followed.

WHO TASK GROUP ON ENVIRONMENTAL HEALTH CRITERIA FOR BENTONITE, KAOLIN, AND SELECTED CLAY MINERALS

A WHO Task Group on Environmental Health Criteria for Bentonite, Kaolin, and Selected Clay Minerals met at Bradford University, Bradford, United Kingdom, from 18 to 22 October 2004. The meeting was opened by Professor Jeffrey Lucas on behalf of the University of Bradford and Dr A. Aitio, Programme for the Promotion of Chemical Safety, WHO, on behalf of the IPCS and its three cooperative organizations (UNEP/ILO/WHO). The Task Group reviewed and revised the draft monograph and made an evaluation of the risks for human health and the environment from exposure to bentonite, kaolin, and other clays.

The first draft was prepared by Dr Zoltán Adamis from the József Fodor National Center for Public Health, National Institute of Chemical Safety, Budapest, Hungary, and Dr Richard B. Williams from the US Environmental Protection Agency, Washington, DC, and Regional Office for the Americas of the World Health Organization. The second draft was also prepared by the same authors in collaboration with the secretariat, which incorporated comments received following the circulation of the first draft to the IPCS contact points for Environmental Health Criteria monographs.

Dr A. Aitio was responsible for the overall scientific content of the monograph.

The efforts of all who helped in the preparation and finalization of the monograph are gratefully acknowledged.

* * *

Risk assessment activities of the International Programme on Chemical Safety are supported financially by the Department of Health and Department for Environment, Food & Rural Affairs, United Kingdom, Environmental Protection Agency, Food and Drug Administration, and National Institute of Environmental Health Sciences, USA, European Commission, German Federal Ministry of Environment, Nature Conservation and Nuclear Safety, Health Canada, Japanese Ministry of Health, Labour and Welfare, and Swiss Agency for Environment, Forests and Landscape.

Task Group Members

Dr Zoltán Adamis, József Fodor National Center for Public Health, National Institute of Chemical Safety, Budapest, Hungary

Prof. Diana Anderson, University of Bradford, Bradford, United Kingdom

Dr Richard L. Attanoos, Llandough Hospital, Cardiff, United Kingdom

Dr Tapan Chakrabarti, National Environmental Engineering Research Institute, Nehru Marg, India

Dr Rogene Henderson (*Co-chair*), Lovelace Respiratory Research Institute, Albuquerque, New Mexico, USA

Dr F. Javier Huertas, CSIC Estacion Experimental del Zaidin, Granada, Spain

Prof. Gunnar Nordberg (*Co-chair*), Umeå University, Umeå, Sweden

Prof. Salah A. Soliman, Alexandria University, Alexandria, Egypt

Prof. Helena Taskinen (*Rapporteur*), Finnish Institute of Occupational Health, Helsinki, Finland

Dr Richard B. Williams, formerly US Environmental Protection Agency, Washington, DC, and Regional Office for the Americas of the World Health Organization

Secretariat

Dr Antero Aitio, International Programme on Chemical Safety, World Health Organization, Geneva, Switzerland

Ms Pearl Harlley, International Programme on Chemical Safety, World Health Organization, Geneva, Switzerland

ACRONYMS AND ABBREVIATIONS

AM	alveolar macrophages
AMAD	activity median aerodynamic diameter
ATP	adenosine triphosphate
CAS	Chemical Abstracts Service
DTA	differential thermal analysis
EDXA	energy-dispersive X-ray analysis
EHC	Environmental Health Criteria monograph
FAO	Food and Agriculture Organization of the United Nations
FEV_1	forced expiratory volume in 1 s
FVC	forced vital capacity
GRAS	generally regarded as safe
IARC	International Agency for Research on Cancer
ILO	International Labour Organization / International Labour Office
IPCS	International Programme on Chemical Safety
IU	international units
JECFA	Joint Expert Committee on Food Additives and Contaminants
JMPR	Joint FAO/WHO Meeting on Pesticide Residues
LC_{50}	median lethal concentration
LD_{50}	median lethal dose
LDH	lactate dehydrogenase
meq	milliequivalent
MMAD	mass median aerodynamic diameter
mppcf	million particles per cubic foot
NIOSH	National Institute for Occupational Safety and Health (USA)
PAM	pulmonary alveolar macrophages
PMN	polymorphonuclear leukocytes
RO	Responsible Officer
SD	standard deviation
TTC	2,3,5-triphenyltetrazolium chloride
UN	United Nations
UNEP	United Nations Environment Programme
USA	United States of America
WHO	World Health Organization

1. SUMMARY

1.1   Identity, physical and chemical properties, and analytical methods

Bentonite is a rock formed of highly colloidal and plastic clays composed mainly of montmorillonite, a clay mineral of the smectite group, and is produced by *in situ* devitrification of volcanic ash. In addition to montmorillonite, bentonite may contain feldspar, cristobalite, and crystalline quartz. The special properties of bentonite are an ability to form thixotrophic gels with water, an ability to absorb large quantities of water, and a high cation exchange capacity. The properties of bentonite are derived from the crystal structure of the smectite group, which is an octahedral alumina sheet between two tetrahedral silica sheets. Variations in interstitial water and exchangeable cations in the interlayer space affect the properties of bentonite and thus the commercial uses of the different types of bentonite. By extension, the term bentonite is applied commercially to any clay with similar properties. Fuller's earth is often a bentonite.

Kaolin or china clay is a mixture of different minerals. Its main component is kaolinite; in addition, it frequently contains quartz, mica, feldspar, illite, and montmorillonite. Kaolinite is made up of tiny sheets of triclinic crystals with pseudohexagonal morphology. It is formed by rock weathering. It has some cation exchange capacity.

The main components of common clay and shale are illite and chlorite. Illite is also a component of ball clays. Illite closely resembles micas, but it has less substitution of aluminium for silicon and/or partial replacement of potassium ions between the unit layers by other cations, such as hydrogen, magnesium, and calcium.

Quantitative measurement of airborne dust containing aluminosilicates is most commonly gravimetric. The methods used for the identification and quantification of aluminosilicates include X-ray diffraction, electron microscopy, energy-dispersive X-ray analysis, differential thermal analysis, electron diffraction, and infrared spectroscopy.

1.2   Sources of human and environmental exposure

Montmorillonite is ubiquitous at low concentrations in soil, in the sediment load of natural waters, and in airborne dust. Biodegradation and bioaccumulation in the food-chain appear minimal, if they occur at all, and abiotic degradation of bentonite into other minerals takes place only on a geological time scale.

Major uses of bentonite include binding foundry sand in moulds; absorbing grease, oil, and animal wastes; pelletizing taconite iron ore; and improving the properties of drilling muds. Speciality uses include serving as an ingredient in ceramics; waterproofing and sealing in civil engineering projects, such as landfill sites and nuclear waste repositories; serving as a filler, stabilizer, or extender in adhesives, paints, cosmetics, and medicines, as a carrier in pesticides and fertilizers, and as a bonding agent in animal feeds; clarifying wine and vegetable oil; and purifying wastewater. The uses of montmorillonite-type Fuller's earth overlap those of bentonite.

Use of kaolin dates back to the third century BC in China. Today it is mined and used in significant quantities for numerous industrial uses. Its most important use is in paper production, where it is used as a coating material. In addition, it is used in great quantities in the paint, rubber, plastic, ceramic, chemical, pharmaceutical, and cosmetics industries.

Common clay and shale, of which illite is often a major component, are used mainly in the manufacture of extruded and other bricks, portland and other cements, concrete blocks and structural concrete, and refractories. Highway surfacing, ceramic tiles, and ceramics and glass are other important uses.

1.3   Environmental levels and human exposure

In view of the widespread distribution of bentonite in nature and its use in an enormous variety of consumer products, general population exposure to low concentrations is ubiquitous.

There is limited information on occupational exposure to bentonite dust in mines, processing plants, and user industries. The highest reported values for total dust and respirable dust concentrations were, respectively, 1430 and 34.9 mg/m^3, although most values were below 10 mg/m^3 for total dust and below 5 mg/m^3 for respirable dust.

Kaolin is a natural component of the soil and occurs widely in ambient air. Kaolin mining and refining involve considerable exposure, and significant exposure is also expected in paper, rubber, and plastic production. Quantitative information on occupational exposure is available for a few countries and industries only. Respirable dust concentrations in kaolin mining and processing are usually below 5 mg/m³.

1.4 Kinetics and metabolism in laboratory animals and humans

No information was available on the kinetics or metabolism of montmorillonite, kaolinite, or illite as they occur in most occupational settings.

Deposition and kinetics of radiolabelled fused montmorillonite after inhalation exposure have been studied in mice, rats, dogs, and humans. Deposition in the nasopharynx increases with particle size and is lower in dogs than in rodents. Tracheobronchial deposition was low and independent of animal species and particle size. Pulmonary deposition was considerably higher in dogs than in rodents and decreased with increasing particle size.

The removal of particles from the lungs took place by solubilization *in situ* and by physical clearance. In dogs, the main mechanism of clearance was solubilization; in rodents, the mechanism of removal was mainly physical transportation. The clearance by mechanical removal was slow, especially in dogs: the half-time was initially 140 days and increased to 6900 days by day 200 post-exposure.

In humans, there was a rapid initial clearance of 8% and 40% of aluminosilicate particles that were, respectively, 1.9 and 6.1 μm in aerodynamic diameter from the lung region over 6 days. Thereafter, 4% and 11% of the two particle sizes were removed following a half-time of 20 days, and the rest with half-times of 330 and 420 days.

Ultrafine particles (<100 nm) have a high deposition in the nasal area; they can penetrate the alveolar/capillary barrier.

1.5 Effects on laboratory mammals and *in vitro* test systems

An important determinant of the toxicity of clays is the content of quartz. The presence of quartz in the clays studied hampers reliable independent estimation of the fibrogenicity of other components of clays.

Single intratracheal injection into rodents of bentonite and montmorillonite with low content of quartz produced dose- and particle size-dependent cytotoxic effects, as well as transient local inflammation, the signs of which included oedema and, consequently, increased lung weight. Single intratracheal exposures of rats to bentonite produced storage foci in the lungs 3–12 months later. After intratracheal exposure of rats to bentonite with a high quartz content, fibrosis was also observed. Bentonite increased the susceptibility of mice to pulmonary infection.

There are limited data on the effects of multiple exposures of experimental animals to montmorillonite or bentonite. Mice maintained on diets containing 10% or 25% bentonite but otherwise adequate to support normal growth displayed slightly reduced growth rates, whereas mice maintained on a similar diet with 50% bentonite showed minimal growth and developed fatty livers and eventually fibrosis of the liver and benign hepatomas (see below).

In vitro studies of the effects of bentonite on a variety of mammalian cell types usually indicated a high degree of cytotoxicity. Concentrations below 1.0 mg/ml of bentonite and montmorillonite particles less than 5 μm in diameter caused membrane damage and even cell lysis, as well as functional changes in several types of cells. The velocity and degree of lysis of sheep erythrocytes were dose dependent.

Kaolin instilled intratracheally produces storage foci, foreign body reaction, and diffuse exudative reaction. After high doses of kaolin (containing 8–65% quartz), fibrosis has been described in some studies, whereas at lower kaolin doses, no fibrosis has been observed in the few available studies.

Information is very limited on the toxicity of illite and non-existent on the toxicity of other components of other commercially important clays. Intratracheally instilled illite with unknown quartz content induced alveolar proteinosis, increased lung weight, and caused collagen synthesis. Illite had limited cytotoxicity towards peritoneal macrophages and was haemolytic *in vitro*.

No adequate studies are available on the carcinogenicity of bentonite. In an inhalation study and in a study using intrapleural injection, kaolin did not induce tumours in rats. No studies are available on the genotoxicity of clays.

Single, very limited studies did not demonstrate developmental toxicity in rats after oral exposure to bentonite or kaolin.

1.6   Effects on humans

General population exposure to low concentrations of montmorillonite and kaolinite, the main components of bentonite and kaolin, respectively, and other clay minerals is ubiquitous. There is no information on the possible effects of such low-level exposure.

Long-term occupational exposures to bentonite dust may cause structural and functional damage to the lungs. However, available data are inadequate to conclusively establish a dose–response relationship or even a cause-and-effect relationship due to limited information on period and intensity of exposure and to confounding factors, such as exposure to silica and tobacco smoke.

Long-term exposure to kaolin causes the development of radiologically diagnosed pneumoconiosis in an exposure-related fashion. Clear-cut deterioration of respiratory function and related symptoms have been reported only in cases with prominent radiological findings. The composition of the clay — i.e., quantity and quality of minerals other than kaolinite — is an important determinant of the effects.

Bentonite, kaolin, and other clays often contain quartz, and exposure to quartz is causally related to silicosis and lung cancer. Statistically significant increases in the incidence of or mortality from chronic bronchitis and pulmonary emphysema have been reported after exposure to quartz.

1.7 Effects on other organisms in the laboratory and field

Bentonite and kaolin have low toxicity to aquatic species, a large number of which have been tested.

1.8 Evaluation of human health risks and effects on the environment

From the limited data available from studies on bentonite-exposed persons, retained montmorillonite appears to effect only mild non-specific tissue changes, which are similar to those that have been described in the spectrum of changes of the "small airways mineral dust disease" (nodular peribronchiolar dust accumulations containing refractile material [montmorillonite] in association with limited interstitial fibrosis). In some of the studies, radiological abnormalities have also been reported.

There exist no reported cases of marked diffuse/nodular pulmonary tissue fibrotic reaction to montmorillonite in the absence of free silica. No quantitative estimates of the potency of bentonite to cause adverse pulmonary effects can be derived.

Long-term exposure to kaolin may lead to a relatively benign pneumoconiosis, known as kaolinosis. Deterioration of lung function has been observed only in cases with prominent radiological alterations. Based on data from china clay workers in the United Kingdom, it can be very roughly estimated that kaolin is at least an order of magnitude less potent than quartz.

Bentonite, kaolin, and other clays often contain quartz, which is known to cause silicosis and lung cancer.

No report on local or systemic adverse effects has been identified from the extensive use of bentonite or kaolin in cosmetics.

The biological effects of clay minerals are influenced by their mineral composition and particle size. The decreasing rank order of the potencies of quartz, kaolinite, and montmorillonite to produce lung damage is consistent with their known relative active surface areas and surface chemistry.

Bentonite and kaolin have low toxicity towards aquatic organisms.

2. IDENTITY, PHYSICAL AND CHEMICAL PROPERTIES, AND ANALYTICAL METHODS

2.1 Introduction

Clay is a widely distributed, abundant mineral resource of major industrial importance for an enormous variety of uses (Ampian, 1985). In both value and amount of annual production, it is one of the leading minerals worldwide. In common with many geological terms, the term "clay" is ambiguous and has multiple meanings: a group of fine-grained minerals — i.e., the clay minerals; a particle size (smaller than silt); and a type of rock — i.e., a sedimentary deposit of fine-grained material usually composed largely of clay minerals (Patterson & Murray, 1983; Bates & Jackson, 1987). In the latter definition, clay also includes fine-grained deposits of non-aluminosilicates such as shale and some argillaceous soils.

This Environmental Health Criteria (EHC) monograph deals with the health hazards associated with bentonite, kaolin, and common clay, which are commercially important clay products, as well as the related phyllosilicate minerals montmorillonite, kaolinite, and illite. Fibrous clay minerals, such as sepiolite, attapulgite, and zeolites, are not discussed. For a recent assessment of the health effects of zeolites, see IARC (1997a). As bentonite, kaolin, and common clay may contain varying amounts of silica, a short summary is presented of recent assessments of quartz (IARC, 1997b; IPCS, 2000).

Recently, there has been an increased interest in the toxicity of airborne fine (0.1–2.5 µm) and ultrafine (<0.1 µm) particles. Epidemiological studies have indicated an increase in morbidity and mortality associated with an increase in airborne particulate matter, particularly in the ultrafine size range (Pekkanen et al., 1997; Stölzel et al., 2003). There is little information on how much of total airborne particulate matter is clay dust and what fraction of airborne clay dusts is in the fine or ultrafine mode.

2.2 Identity

2.2.1 *Bentonite*

The term "bentonite" is ambiguous. As defined by geologists, it is a rock formed of highly colloidal and plastic clays composed mainly of montmorillonite, a clay mineral of the smectite group (Fig. 1), and is produced by *in situ* devitrification of volcanic ash (Parker, 1988). The transformation of ash to bentonite apparently takes place only in water (certainly seawater, probably alkaline lakes, and possibly other fresh water) during or after deposition (Grim, 1968; Patterson & Murray, 1983). Bentonite was named after Fort Benton (Wyoming, USA), the locality where it was first found. In addition to montmorillonite, bentonite may also contain feldspar, biotite, kaolinite, illite, cristobalite, pyroxene, zircon, and crystalline quartz (Parkes, 1982).

By extension, the term bentonite is applied commercially to any plastic, colloidal, and swelling clay regardless of its geological origin. Such clays are ordinarily composed largely of minerals of the montmorillonite group.

Bentonite is a rock or a clay base industrial material. It is therefore a mixture of minerals. No "molecular" formula can be given.

The Chemical Abstracts Service (CAS) registry number for bentonite is 1302-78-9.

Synonyms and trade names used to designate bentonite include Albagel Premium USP 4444, Bentonite magma, Bentonite 2073, Bentopharm, CI 77004, E558, HI-Gel, HI-Jel, Imvite I.G.B.A., Magbond, mineral sopa, Montmorillonite, Panther creek bentonite, soap clay, Southern bentonite, taylorite, Tixoton, Veegum HS, Volclay, Volclay Bentonite BC, and Wilkinite (CIREP, 2003; RTECS, 2003a).

The term "montmorillonite" is also ambiguous and is used both for a group of related clay minerals and for a specific member of that group (Bates & Jackson, 1987). For the former use, smectite is more appropriate (a group of clay minerals that includes montmorillonite, saponite, sauconite, beidellite, nontronite, etc.) (Fig. 1).

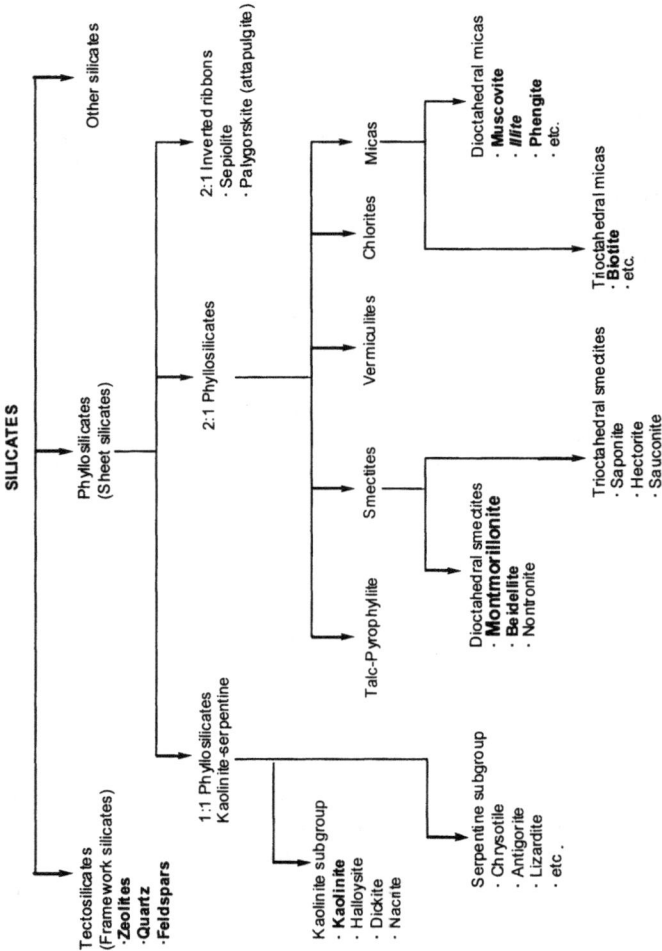

Fig. 1. Classification of silicates (Bailey, 1980b; Rieder et al., 1998). Minerals that can be frequently found in bentonite or kaolin are in bold; the main components are in large typeface. *Illite* is a component of common soil and sediments and is classified as a mica.

The basic crystal structure of smectites is an octahedral alumina sheet between two tetrahedral silica sheets. Atoms in these sheets common to both layers are oxygens. These three-layer units are stacked one above another with oxygens in neighbouring layers adjacent to each other. This produces a weak bond, allowing water and other polar molecules to enter between layers and induce an expansion of the mineral structure. In the tetrahedral coordination, silicon may be substituted by aluminium and possibly phosphorus; in the octahedral coordination, aluminium may be substituted by magnesium, iron, lithium, chromium, zinc, or nickel. Differences in the substitutions within the lattice in terms of position and elemental composition give rise to the various montmorillonite clay minerals: montmorillonite, nontronite, saponite, hectorite, sauconite, beidellite, volkhonskoite, pimelite, and griffithite.

The molecular formula for montmorillonite is usually given as $(M^+_x \cdot nH_2O)$ $(Al_{2-y}Mg_x)Si_4O_{10}(OH)_2$, where $M^+ = Na^+$, K^+, Mg^{2+}, or Ca^{2+} (Brindley & Brown, 1980). Ideally, $x = 0.33$. The CAS registry number for montmorillonite is 1318-93-0.

Fuller's earth is loosely defined by its properties and commercial use and not by its mineralogy and molecular structure (Grim, 1968; Patterson & Murray, 1983; Ampian, 1985; Bates & Jackson, 1987; Hosterman & Patterson, 1992). The term is derived from the initial usage of certain clays in bleaching, degreasing, and fulling (shrinking) woollen fabrics. Grim (1968) defines Fuller's earth as "any natural earthy material which will decolorize mineral or vegetable oils to a sufficient extent to be of economic importance." The term encompasses not only some bentonites but also some kaolinites and some silts with minimal clay content. In the USA, the term Fuller's earth is used to mean palygorskite, a fibrous magnesium–aluminium silicate mineral close to sepiolite (IARC, 1997c). Data on production and other aspects of Fuller's earth often do not distinguish between bentonite and non-bentonite types.

2.2.2   Kaolin

The name "kaolin" is derived from the word Kau-Ling, or high ridge, the name given to a hill near Jau-chau Fu, China, where kaolin was first mined (Sepulveda et al., 1983). Kaolin, commonly referred to as china clay, is a clay that contains 10–95% of the mineral kaolinite and usually consists mainly of kaolinite (85–95%). In addition to kaolinite, kaolin usually contains quartz and mica and also, less frequently, feldspar, illite, montmorillonite, ilmenite, anastase, haematite, bauxite, zircon, rutile, kyanite, silliminate, graphite, attapulgite, and halloysite. Some clays used for purposes similar to those for which kaolin is used may contain substantial amounts of quartz: "kaolin-like" clays used in South African pottery contained 23–58% quartz and, as the other major constituent, 20–36% kaolinite (Rees et al., 1992).

The CAS registry number for kaolin is 1332-58-7.

Synonyms and trade names for kaolin include Altowhites, Argilla, Bentone, Bolbus alba, China clay, CI 77004, Emathlite, Fitrol, Fitrol desiccite 25, Glomax, Hydrite, Kaopaous, Kaophills-2, Kolite, Langford, McNamee, Parclay, Pigment white 19, Porcelain clay, Snow tex, terra alba, and white bole (CIREP, 2003; RTECS, 2003b).

The structure of kaolinite is a tetrahedral silica sheet alternating with an octahedral alumina sheet. These sheets are arranged so that the tips of the silica tetrahedrons and the adjacent layers of the octahedral sheet form a common layer (Grim, 1968). In the layer common to the octahedral and tetrahedral groups, two-thirds of the oxygen atoms are shared by the silicon and aluminium, and then they become O instead of OH. The charges within the structural unit are balanced. Analyses of many samples of kaolinite minerals have shown that there is very little substitution in the lattice (Grim, 1968). The molecular formula that is common for the kaolinite group (kaolinite, nacrite, dickite) is $Al_2Si_2O_5(OH)_4$ (Grim, 1968).

The CAS registry number for kaolinite is 1318-74-7.

2.2.3 Other clays

In addition to bentonite (and Fuller's earth) and kaolin, the main types of commercial clays are ball clay, common clay and shale, and fire clay. Ball clay consists primarily of kaolinite, with minor amounts of illite, chlorite, smectite minerals, quartz, and organic materials. Common clay and shale contain illite and chlorite as major components, while fire clay comprises mainly kaolinite, halloysite, and/or diaspore (Virta, 2002).

Illite is a name for a group of mica-like clay minerals proposed by R.E. Grim, R.H. Bray, and W.F. Bradley in 1937 to honour the US state of Illinois, a source of illite and a state that had supported research on clay (Grim, 1968; Bates & Jackson, 1987). Illite clay minerals show substantially no expanding lattice characteristics and are characterized by intense 1.0-nm 001 and 0.33-nm 003 peaks that are not modified by glycerol or ethylene glycol solvation, potassium saturation, or heating to 550 °C (Fanning et al., 1989). Grim (1968) stated that illite includes both trioctahedral (biotite) and dioctahedral (muscovite) types of crystallization. The US Geological Survey report entitled *A Laboratory Manual for X-Ray Powder Diffraction*, however, limits illite to minerals with a dioctahedral, muscovite structure (USGS, 2001). Velde (1995) separated mica-like minerals such as illite from true micas on the basis that the former have almost exclusively potassium as the interlayer ion and a charge imbalance always slightly less than 1.0 per unit cell. (This value for the charge imbalance is much below the 1.3–1.5 given by Grim [1968].) In addition, true micas usually have a greater grain size than the mica-like clay minerals (i.e., >2 μm) and commonly are found in rocks that have been subjected to elevated temperatures.

The Association Internationale Pour l'Étude des Argiles Nomenclature Committee (Rieder et al., 1998) defines illite as a serial name for dioctahedral interlayer-deficient mica with composition $K_{0.65}(Al_2)(Si_{3.35}Al_{0.65})O_{10}(OH)_2$. The aluminium in octahedral positions can be partially substituted by Mg^{2+}, Fe^{2+}, and Fe^{3+}.

The CAS registry number of illite is 1217-36-03 (Grim, 1968). Synonyms of illite include hydromica, illidromica, and glimmerton.

The chemical composition of illite deposits from Malaysia is summarized in Fan & Aw (1989) and from the USA and Germany in Grim (1968) and Gaudette et al. (1966). The water content of Malaysian illite ranged from 19% to 26%, and that of the illites described by Grim (1968), from 6% to 12%. The density of illite is 2.6–2.9 g/cm^3 (Grim, 1968; Tamás, 1982). Gaudette et al. (1966) mentioned that illites from Illinois, Oklahoma, and Wisconsin, USA, were light green or light to dark grey clays. Illite mined in Füzérradvány, Hungary, is described as greasy, clay-like, and of high plasticity (Tamás, 1982). Its colour is determined by its iron content and may be white, yellow, lilac, or reddish brown.

2.3 Physical and chemical properties

2.3.1 Bentonite

Bentonite feels greasy and soap-like to the touch (Bates & Jackson, 1987). Freshly exposed bentonite is white to pale green or blue and, with exposure, darkens in time to yellow, red, or brown (Parker, 1988). The special properties of bentonite are an ability to form thixotrophic gels with water, an ability to absorb large quantities of water with an accompanying increase in volume of as much as 12–15 times its dry bulk, and a high cation exchange capacity.

Substitutions of silicon by cations produce an excess of negative charges in the lattice, which is balanced by cations (Na^+, K^+, Mg^{2+}, Ca^{2+}) in the interlayer space. These cations are exchangeable due to their loose binding and, together with broken bonds (approximately 20% of exchange capacity), give montmorillonite a rather high (about 100 meq/100 g) cation exchange capacity, which is little affected by particle size. This cation exchange capacity allows the mineral to bind not only inorganic cations such as caesium but also organic cations such as the herbicides diquat, paraquat (Weber et al., 1965), and s-triazines (Weber, 1970), and even bio-organic particles such as rheo-viruses (Lipson & Stotzky, 1983) and proteins (Potter & Stollerman, 1961), which appear to act as cations. Variation in exchangeable cations affects the maximum amount of water uptake and swelling. These are greatest with sodium and least with potassium and magnesium.

Interstitial water held in the clay mineral lattice is an additional major factor controlling the plastic, bonding, compaction, suspension, and other properties of montmorillonite-group clay minerals. Within each crystal, the water layer appears to be an integral number of molecules in thickness. Physical characteristics of bentonite are affected by whether the montmorillonite composing it has water layers of uniform thickness or whether it is a mixture of hydrates with water layers of more than one thickness. Loss of absorbed water from between the silicate sheets takes place at relatively low temperatures (100–200 °C). Loss of structural water (i.e., the hydroxyls) begins at 450–500 °C and is complete at 600–750 °C. Further heating to 800–900 °C disintegrates the crystal lattice and produces a variety of phases, such as mullite, cristobalite, and cordierite, depending on initial composition and structure. The ability of montmorillonite to rapidly take up water and expand is lost after heating to a critical temperature, which ranges from 105 to 390 °C, depending on the composition of the exchangeable cations. The ability to take up water affects the utilization and commercial value of bentonite (Grim, 1968; Parker, 1988).

Montmorillonite clay minerals occur as minute particles, which, under electron microscopy, appear as aggregates of irregular or hexagonal flakes or, less commonly, of thin laths (Grim, 1968). Differences in substitution affect and in some cases control morphology.

2.3.2   Kaolin

Kaolinite, the main constituent of kaolin, is formed by rock weathering. It is white, greyish-white, or slightly coloured. It is made up of tiny, thin, pseudohexagonal, flexible sheets of triclinic crystal with a diameter of 0.2–12 µm. It has a density of 2.1–2.6 g/cm^3. The cation exchange capacity of kaolinite is considerably less than that of montmorillonite, in the order of 2–10 meq/100 g, depending on the particle size, but the rate of the exchange reaction is rapid, almost instantaneous (Grim, 1968). Kaolinite adsorbs small molecular substances such as lecithin, quinoline, paraquat, and diquat, but also proteins, polyacrylonitrile, bacteria, and viruses (McLaren et al., 1958; Mortensen, 1961; Weber et al., 1965; Steel & Anderson, 1972; Wallace et al., 1975; Adamis & Timár, 1980; Schiffenbauer & Stotzky, 1982; Lipson & Stotzky, 1983). The adsorbed material can be easily removed from the particles because adsorption is limited to the surface of the particles (planes, edges), unlike the case with montmorillonite, where the adsorbed molecules are also bound between the layers (Weber et al., 1965).

Upon heating, kaolinite starts to lose water at approximately 400 °C, and the dehydration approaches completeness at approximately 525 °C (Grim, 1968). The dehydration depends on the particle size and crystallinity.

2.3.3   Other clays

Illite, together with chlorite, is the main component of common clay and shale. It is also an important impurity in limestone, which can affect the properties and thus the value of the stone for construction and other purposes (Carr et al., 1994). Despite the widespread occurrence of illite in nature, large deposits of high purity are quite rare (as discussed in section 3.2.1.3).

Illite usually occurs as very small (0.1–2 µm), poorly defined flakes commonly grouped into irregular aggregates. Lath-shaped and ribbon-shaped illite particles up to 30 µm in length and 0.1–0.3 µm in width have also been described (Srodon & Eberl, 1984), but their existence is controversial. Velde (1985) stated unqualifiedly that these so-called filamentous illites are mixed-layer structures. Srodon & Eberl (1984), however, drawing on the same references plus their own data, concluded that these filaments in some cases are mixed-layer structures but in other cases are composed only of illite, and they further supported their view with scanning electron microscopic photographs of lath-shaped crystals of what they identified as illite.

The special properties of illite are derived from its molecular structure. The balancing cation is mainly or entirely potassium, and charge deficiency from substitutions is at least twice that of smectites (i.e., 1.3–1.5 per unit cell layer) and is mainly in the silica sheet and close to the surface of the unit layer rather than in the octahedral layer as in smectites (Grim, 1968). These differences from smectites produce a structure in which interlayer balancing cations are not easily exchanged and the unit layers are relatively fixed in position and do not permit polar ions such as water to readily enter between them and produce expansion.

Illite reacts with both inorganic and organic ions and has a cation exchange capacity of 10–40 meq/100 g, a value intermediate between those of montmorillonite and kaolinite (Grim, 1968). Ion exchange capacity is reduced by heating. The potassium in the interlayer space is "fixed" to a considerable degree, making it not readily available to plants, a matter of importance in soil science and agriculture. A portion of the interlayer potassium, however, can be slowly leached, leading to the degradation of the illite. Such degradation, however, can be reversed by the addition of potassium. Wilken & Wirth (1986) stated that fithian illite from Illinois, USA, adsorbed hexachlorobenzene suspended in distilled water with a sorption partition coefficient of 2200–2600 and that more than half of this adsorbed hexachlorobenzene could be desorbed by further contact with distilled water. However, the fithian illite used in the experiment had a composition of 30% quartz, 19% feldspar, 11% kaolinite, 1% organic carbon, and 40% illite, making it impossible to know how much of the measured adsorption could be ascribed to illite.

The dehydration and other changes in illite with heating have been studied by several investigators, with inconsistent results (Grim, 1968). Some of the inconsistency in findings may result from differences in the period at which samples were held at a given temperature, since dehydration is a function of both time and temperature (Roy, 1949). It is also probable that small differences in particle size, crystal structure, and molecular composition among samples of what were ostensibly the same mineral contributed to the inconsistencies. Dehydration takes place either smoothly or in steps between about 100 and 800 or 850 °C for both biotite and muscovite illites. Loss of structure by the various illite minerals occurs between about 850 and 1000 °C.

2.3.4   Surface chemistry

Chemical properties of the surfaces of silicates and, in particular, clay minerals are strongly dependent on their mineral structure. The basic unit of montmorillonite crystals is an extended layer composed of an octahedral alumina sheet (O) between two tetrahedral silica sheets (T), forming a TOT unit (Bailey, 1980a). The stacks of TOT layers produce the montmorillonite crystals. Isomorphic substitutions in the octahedral sheet (few tetrahedral substitutions are observed in montmorillonite) create an excess of negative structural charge that is delocalized in the lattice. Cations located between two consecutive layers contribute to compensate the structural charge and to keep the layers bound. These cations can easily be exchanged, since they are retained by electrostatic attractions. The surface area associated with the basal surfaces of the extended TOT units is known as the interlayer surface when it corresponds to consecutive layers or as the external surface when it corresponds to the external basal surfaces of a crystal. The external and interlayer surface area represents approximately 95% of the total surface area of montmorillonite. On the other hand, the periodic structure of the montmorillonite crystals is interrupted at the edges, where the broken bonds compensate their charge by the specific adsorption of protons and water molecules (Schindler & Stumm, 1987; Stumm & Wollast, 1990; Stumm, 1997). This interruption of the periodic structure confers to the edge surface an amphoteric character — i.e., a pH-dependent surface charge and the capacity to react speci-fically with cations, anions, and molecules (organic and inorganic), forming chemical bonds.

The kaolinite layers are formed only by a tetrahedral sheet of silica and an octahedral sheet of alumina, which contain almost no isomorphic substitutions (Bailey, 1980a). Hydroxyls of the octahedral sheet bind to oxygens of the tetrahedral sheet of the consecutive layer in the kaolinite crystal. The structural charge of the crystal is non-existent, and no molecules or cations are present in the interlayer space. The surface area of the kaolinite is thus reduced to external surface area and to edge surface area. The edge surface area is similar in nature and properties to the one observed in montmorillonite and represents approximately 20% of the total surface area. However, two types of external surfaces are defined for kaolinite: one associated with the outermost tetrahedral sheet and the other associated with the outermost octahedral sheet. The reactivity of the external octahedral sheet is due to the hydroxyl groups present, which i) can produce a pH-dependent surface charge by protonation and deprotonation reactions and ii) can react specifically with other molecules (Huertas et al., 1998).

Quartz is the most relevant accessory mineral in bentonite and kaolin. Its structure is a tridimensional framework of silicate tetrahedra arranged sharing corners. In contrast with montmorillonite and kaolinite, quartz surface area is defined only as an external surface. Broken bonds, which compensate charge by adsorption of protons or water molecules, are found in all the crystal phases.

The "active" surface area is close to 100% in quartz, 20% in kaolinite, and 5% in montmorillonite. This may be related to the observed toxicity of these materials (see chapter 6).

2.3.5 *Trace elements in clays*

Knowledge of mechanisms controlling the distribution of trace elements in clays is scarce and contradictory, in spite of the many investigations carried out on the geochemical behaviour of some trace elements used in geological reconstruction as "geochemical indicators" (Fiore et al., 2003). It is well known that clays contain trace elements that literature indicate as toxic and/or micronutrients (i.e., antimony, arsenic, cadmium, cobalt, copper, lead, mercury, nickel, selenium, tellurium, thallium, zinc) whose concentrations are widely variable, depending on their geological history. These trace elements may be in the clay (or accessory) mineral structure as well as adsorbed on clay particles, which play the most important role in controlling their distribution and abundance. Chemical elements in crystalline positions are usually "locked," whereas those adsorbed may be mobilized and transferred to leaching solutions.

For information on the leaching and bioavailability of clay components, see chapter 5.

2.4 Analytical methods

2.4.1 *Quantitative measurement of dust*

Inhaled dust appears to be the major mode of human exposure to aluminosilicates, and a general review of methodology for dust sampling and analysis is presented in Degueldre (1983). Quantitative methods for the measurement of total dust and respirable dust are described in the *NIOSH Manual of Analytical Methods* (Eller & Cassinelli, 1994), in Part 40 of the US government's Code of Federal Regulations (US EPA, 1996a), and in *Mine Safety and Health Administration Handbook No. PH90-IV-4* (US Department of Labor, 1990). All of these methods are gravimetric and measure the weight of dust collected on a filter from a known volume of air.

Although the basic principles behind these methods are simple, accurate and precise results demand sensitive equipment and attention to detail. The National Institute for Occupational Safety and Health (NIOSH) method for total dust requires an appropriate sampler with a polyvinyl chloride or equivalent filter with a pore size of 2–5 μm, a pump with a well characterized, uniform rate of air flow of 1.5–2.0 litres/min, and a microbalance capable of weighing to 0.01 mg, as well as a vacuum desiccator, a static neutralizer, and a chamber with constant temperature and humidity for storing the filters. Guidance for optimizing performance of the 10-mm cyclone sampler is provided by Bartley & Breuer (1982). Briant & Moss (1984) recommended use of a conducting, graphite-filled nylon cyclone to avoid developing an electrostatic charge on the cyclone, which may reduce the measured value for respirable dust below the actual value. Significant errors can also be introduced by air leakage in the filter assembly, the accumulation of static electrical charge on the filter, and many other factors.

Another quantitative measurement of dusts is the determination of particle number. The toxicity of particles has been shown to be related more to surface than to mass. The smaller the particle, the larger the surface area to mass ratio.

2.4.2 Identification of phyllosilicates

There is no single or simple procedure for the positive identification of montmorillonite-group or other aluminosilicates or for their quantification in dust and other samples. The application of several methods may be necessary for even approximate identification and rough quantification. These methods include X-ray diffraction, electron microscopy, energy-dispersive X-ray analysis, differential thermal analysis, and infrared spectroscopy. In the past, chemical methods based on differences in resistance of various clay minerals to chemical attack, the so-called "rational methods of analysis," were used.

X-ray powder diffraction analysis is the basic technique for clay mineral analysis (Moore & Reynolds, 1989). After preliminary removal of sand, clay is separated from silt by centrifugation or sedimentation from suspensions. X-ray diffraction patterns are obtained for air-dried samples and, in the case of oriented aggregates, also for samples treated with ethylene glycol vapour or heated to 350 and 550 °C. Diffraction patterns are compared with standards (Grim, 1968; Thorez, 1975; Brindley & Brown, 1980; JCPDS, 1981) for identification of minerals. Comparisons are complicated, however, by variations in diffraction patterns arising from differences in amounts of absorbed water, by the presence of imperfections in the crystal lattice structure of the minerals, and by mixed-layer structures formed by interstratification of minerals within a single particle (Grim, 1968). Approximate quantification of mineral abundance in samples containing several minerals is possible, although subject to a variety of complications and errors (Starkey et al., 1984; Salt, 1985).

Transmission electron microscopy is valuable for identifying aluminosilicates with a distinctive morphology (Starkey et al., 1984). Particles dispersed on a plastic film can be observed directly by transmission microscopy or shadowed to increase contrast by evaporating a heavy metal onto specimens prior to examination. Grim (1968) and Starkey et al. (1984) provide an entry into the extensive literature on this subject. Many aluminosilicates (e.g., montmorillonite, which occurs as broad mosaic sheets decomposing into minute flakes) lack a distinctive morphology and cannot be identified by this technique.

Energy-dispersive X-ray analysis (EDXA) — also referred to as energy-dispersive X-ray microanalysis, X-ray microanalysis, electron microscopic microanalysis, and energy-dispersive X-ray spectrometry — and electron diffraction may permit the rapid identification of individual clay mineral particles (Sahle et al., 1990; Lee, 1993) and have been applied particularly to the identification of inhaled particles sampled via bronchoalveolar lavage or from lung specimens (Johnson et al., 1986; Costabel et al., 1990; Monsó et al., 1991, 1997; Chariot et al., 1992; Bernstein et al., 1994; Dufresne et al., 1994). EDXA requires a scanning or transmission electron microscope equipped with an energy-dispersive X-ray spectrometer and appropriate mathematical tools for analysing the resulting spectra. EDXA identifies and quantifies elements above atomic number 8. Since the basic classification of clay minerals is based on structural formula and the atomic composition is similar for many different clay minerals, EDXA cannot provide secure identification except by comparison with standards previously identified by other means. Application of EDXA without appropriate standards is likely to generate significant errors (Newbury et al., 1995). In practice, EDXA is ordinarily combined with conventional transmission electron microscopy to first visualize a particle. Probe size is then adjusted downward so that only the selected particle is analysed. The best results are obtained by operating the microanalysis in scanning transmission mode.

Differential thermal analysis (DTA) is based on temperature differences between the sample and a thermally inert material during heating or cooling and is most useful for mineral identification in samples composed mainly or entirely of a single clay mineral (Grim & Rowland, 1944; Mackenzie, 1970; Smykatz-Kloss, 1974). The sample and a control, a thermally inert material (e.g., aluminium oxide), are heated in separate crucibles in a well regulated furnace. During heating, the change in temperature of the clay sample will be modified by endothermic and exothermic reactions. Identification is done by comparing the temperature difference trace with standard curves (Grim & Rowland, 1944; Mackenzie, 1970; Smykatz-Kloss, 1974). DTA is not without difficulties, since results are influenced by furnace atmosphere, heat conductance of sample and crucible, type of thermocouple, rate of heating, grain size of sample, and many other factors (Smykatz-Kloss, 1974). Identification of component minerals in mixtures and quantification of mixtures by use of DTA are difficult and often may be impossible due to overlapping DTA curves (Grim, 1947; Smykatz-Kloss, 1974).

Many aluminosilicates can be identified by distinctive infrared absorption spectra. Absorption is normally measured over a range of frequencies (Grim, 1968), and the resulting spectrum compared with published standards (Van der Marel & Beutelspacher, 1976; Ferraro,

1982). Satisfactory measurements require appropriate mounting of specimens and minimizing of scattering and reflection. The latter is accomplished by using particles smaller than the minimum wavelength (Grim, 1968).

Montmorillonite can be distinguished from beidellite, nontronite, and saponite by its irreversible collapse of the structure after Li saturation and heat treatment. Li migration to the octahedral charge converts montmorillonite to a non-expandable structure upon treatment with water, glycol, or ethylene glycol. The procedure is known as the Greene-Kelly test (Greene-Kelly, 1955). Other smectites, such as beidellite and saponite, should expand to give the characteristic 1.77-nm reflection upon solvation with ethylene glycol.

3. SOURCES OF HUMAN AND ENVIRONMENTAL EXPOSURE

3.1 Natural occurrence

3.1.1 Bentonite

Bentonite derived from ash falls tends to be in beds of uniform thickness (from a few millimetres to 15 m) and extensive over large areas (Parker, 1988). Bentonite from ash falls and other sources occurs worldwide in strata spanning a broad range of ages, but is most abundant in Cretaceous or younger rocks.

Bentonite is a widely distributed material. Accordingly, its major component, montmorillonite, occurs abundantly as dust at and near surface deposits of bentonite and is dispersed widely by air and moving water. Montmorillonite is thus ubiquitous in low concentrations worldwide in soil, in the sediment load of natural waters, and in airborne dust. Biodegradation appears minimal, if it occurs at all, and there is no evidence of or reason to suspect accumulation in the foodchain. Abiotic degradation of bentonite into other minerals takes place only on a geological time scale (Parker, 1988).

3.1.2 Kaolin

Kaolin and the clay mineral kaolinite are natural components of the soil and occur widely in ambient air as floating dust. Kaolinite is formed mainly by decomposition of feldspars (potassium feldspars), granite, and aluminium silicates. It is also not uncommon to find kaolin deposited together with other minerals (illite, bentonite). The process of kaolin formation is called kaolinization (Grim, 1968).

Kaolinite formation occurs in three ways:

- crumbling and transformation of rocks due to the effects of climatic factors (Zettlitz type);
- transformation of rocks due to hydrothermal effects (Cornwall type); and
- formation by climatic and hydrothermal effects (mixed type).

The type of clay mineral formed during the decay of rocks containing aluminium silicates is influenced by the climate, the aluminium/silicon ratio, and pH. Conditions conducive for kaolinite formation are strong dissolution of Ca^{2+}, Mg^{2+}, and K^+ ions and the presence of H^+ ions (pH 4–5) (Parker, 1988).

Kaolinite quarries can be categorized according to the geohistorical age of the parent rock:

- Precambrian (Ukraine, Spain, Czech and Slovak republics)
- Postcambrian (Cornwall, England, and Ural Mountains, Russia)
- Palaeovolcanite (Meissen, Germany)
- Neovolcanite (Tokaj Mountain, Hungary).

Kaolinite can also be categorized according to whether it remained at the place of formation or was transported (Parkes, 1982; Kuzvart, 1984):

- *primary*: lateritic types formed mainly under tropical climatic conditions (South America, Africa, Australia); and
- *secondary*: transported by different forces, water, wind (Georgia, USA).

Owing to the different ways in which kaolin can form, several kinds of minerals may occur in natural kaolins. For example, the kaolin of Cornwall, England, contains 10–40% kaolinite; the rest is made of quartz, mica, and feldspar. The kaolin of Georgia, USA, contains 85–95% kaolinite, as well as quartz, muscovite, and feldspar (Patterson & Murray, 1975).

3.1.3 Other clays

Illite is widely distributed in nature, abundant, and often the dominant clay mineral in soil, terrestrial deposits, sedimentary rocks, freshwater sediments, and most deep-sea clays (Grim, 1968). It is also an important impurity in limestone, affecting the properties and thus the value of the stone for construction and other purposes (Carr et al., 1994). Despite the widespread occurrence of illite in nature, large deposits of high purity are quite rare.

Illite, together with smectites, is formed by weathering of acid igneous rock containing considerable quantities of potassium and magnesium under conditions that permit the potash and magnesia to remain in the weathering environment after breakdown of the parent material (Grim, 1968). Deer et al. (1975) and Srodon & Eberl (1984) identified many potential sources of illite, including weathering of silicates (primarily feldspar), alteration of other clay minerals, and degradation of muscovite. Righi & Meunier (1995) concluded that illite can also be degraded by weathering into illite–montmorillonite (smectite) mixed-layer minerals.

3.2 Anthropogenic sources

3.2.1 Production levels and sources

3.2.1.1 Bentonite

Bentonite (Table 1) and bentonite Fuller's earth (Table 2) are mined worldwide. The USA is the major producer of bentonite Fuller's earth (Table 2). Approximately 90% of world bentonite production is concentrated in 13 countries: the USA, Greece, the Commonwealth of Independent States, Turkey, Germany, Italy, Japan, Mexico, Ukraine, Bulgaria, Czech Republic, South Africa, and Australia (Table 1). The USA, Greece, and the Commonwealth of Independent States account for roughly 55% of the annual world production of 10 million tonnes. Wyoming produces the bulk of bentonite mined in the USA (Ampian, 1985). In 2002, the bulk of US production was used domestically, and only a small fraction, 11%, was exported worldwide (Virta, 2002). In addition to the mining of natural deposits, small amounts of bentonite, mainly hectorite, are produced synthetically in both Europe and the USA for use as a catalyst.

Most bentonite is mined by stripping methods from open pits after removing any overburden, although underground methods are used in a few places, such as the Combe Hay district in the United Kingdom (Patterson & Murray, 1983). Since deposits are often not uniform in composition, bentonite from a single pit may be separated into several stockpiles, which subsequently are blended to obtain the desired composition. Bentonite is usually processed by breaking large pieces into smaller fragments, drying at low to moderate temperatures to remove water and other volatiles without altering the molecular structure of the

Table 1. Bentonite production in selected countries[a–d]

Country	Bentonite production (kilotonnes)[e]			
	1991	1995	1999	2002[f]
Algeria[g]	26	17	15	27[h]
Argentina	108	111	129	89[h]
Australia[f,g]	35	35	180	200[g]
Brazil, beneficiated	130	150	275	165
Bulgaria	NA	126	243	250
Commonwealth of Independent States[f,i]	2 400	1 300	700	750
Croatia	XX	7	8	11
Cyprus	59	50	139	125
Czech Republic	XX	54	160	250
Egypt[f]	6[h]	2[h]	50	50
Germany[f]	583[h]	529[h]	500[h]	500
Greece[f]	600[h]	1 115[h]	950[h]	950
Hungary	18	23	16	30
Iran[j]	40	55	65	80
Italy	402	591	500	500
Japan	554	478	428	400
Macedonia[f]	XX	30	30	30
Mexico	145	73	209	400
Morocco	9	29	37	66[h]
Pakistan	5	6	15	28
Peru	15	27	20	18
Poland	35	6	5[k]	30[k]
Romania	150[f]	42	20	15
South Africa[l]	65	71	50	218[h]
Spain[f]	150	173	150[h]	150
Turkey	124	602	900	600
Turkmenistan[f]	XX	50	50	50
Ukraine[f]	XX	NA	300	300
USA	3 430	3 820	4 070	3 970[h]
Total	9 360	9 800	10 400	10 300

Table 1 (Contd)

[a] From USGS (1997); Virta (2001, 2002).
[b] Figures are rounded to the nearest 1000 tonnes and may not add to totals shown. Listed are countries producing >10 000 tonnes each (approximately 0.1% of the world total production) in 2002.
[c] Tables include data available through August 21, 2003.
[d] In addition to the countries listed, Canada and China are believed to produce bentonite, but output is not reported, and available information is inadequate to make reliable estimates of output levels.
[e] NA = not available; XX = not applicable.
[f] Figures are estimated, unless otherwise indicated (see footnote h).
[g] Includes bentonitic clays.
[h] Reported figure.
[i] Does not include Armenia, Georgia, or Turkmenistan for the year 1995 or Armenia, Georgia, Turkmenistan, or Ukraine for the years 1999 and 2002.
[j] Year beginning March 21 of that stated
[k] Montmorillonite-type bleaching clay.
[l] May include other clays.

bentonite, and grinding to the desired size. The desired size is generally 200 mesh (US standard sieve size) or finer, which is equivalent to particle diameters of less than 70 µm. A coarser granular material is also produced for kitty litter applications. Processing may also include beneficiation, which may involve removing sand and other impurities as well as modifying the type of exchangeable ions in the crystal lattice. Some of the calcium bentonite produced in Texas, USA, is, for example, treated with sodium hydroxide to replace calcium with sodium in the montmorillonite and make the resulting product more suitable for use in drilling mud (Hosterman & Patterson, 1992). An organic-clad bentonite is produced for speciality purposes (paint, speciality greases, etc.) by replacing the inorganic exchangeable cations in the montmorillonite with an alkyl ammonium organic cation and is marketed under trade names such as Bentone and Nikkagel.

3.2.1.2 Kaolin

Large quantities of kaolin are mined and traded internationally. Virta (2002) lists 55 countries with a production of more than 1000 tonnes per year. Production figures for each country that produced >0.1% (>43 000 tonnes) of the estimated world production of kaolin in 2002 are given in Table 3. The People's Republic of China was estimated to produce 1.9, 2, and 2.12 million tonnes in 2000, 2001, and 2002 (Lines, 2003). The estimated annual production capacity of kaolin in China was 3.2–3.4 million tonnes in 2002–2003, out of which washed kaolin was only 700 000 tonnes (Ma & Tang, 2002; Moore, 2003).

Table 2. World bentonite Fuller's earth production[a,b]

Country	Fuller's earth production (kilotonnes)			
	1991	1995	1999	2002[c]
Algeria	5	5[c]	3	4[d]
Argentina[c]	2	2	2	2
Australia (attapulgite)	15[c]	15[c]	6	6
Germany (unprocessed)[c]	708[d]	600	500	500
Italy[c]	23[d]	34[d]	30	30
Mexico	41	16	48	150
Morocco (smectite)	38	15	22	42[d]
Pakistan	22	13	16	15
Senegal (attapulgite)	129	120	136	176[d]
South Africa (attapulgite)[c]	8	8	7	8
Spain (attapulgite)[c]	73	94	90	90
United Kingdom[c,e]	189[d]	132	140	140
USA[f]	2740	2640	2560	2730
Total	3990	3690	3560	3890

[a] From USGS (1997); Virta (2001, 2002).
[b] Figures are rounded to the nearest 1000 tonnes. Table excludes former and current centrally planned economy countries, some of which may produce Fuller's earth, but for which no data are available. In addition to the countries listed, France, India, Iran, Japan, and Turkey reportedly have produced Fuller's earth in the past and may continue to do so, but no data are available.
[c] Figures are estimated, unless otherwise indicated (see footnote d).
[d] Reported figure.
[e] Saleable product.
[f] Sold or used by producers.

Kaolin is usually removed from the mines in large moist lumps, and the initial process of refining involves chiefly a change in the physical state. Two main methods have been used for the refining of the natural clay: the air flotation or dry process, and the wet process. In the dry process, the clay is dried, pulverized, and then carried by currents of hot air into classifying chambers, from which it emerges as a stream of finely powdered kaolin. This process, which generates huge amounts of dust, was generally used prior to 1940 and has now been largely replaced by the wet process. In this process, the crude clay is mixed with water and vigorously agitated, and the individual kaolin particles are thus separated from each other and suspended in water. Water is then removed using different procedures (Edenfield, 1960).

Table 3. Kaolin production in selected countries[a,b]

Country	Kaolin production (kilotonnes)			
	1991	1995	1999	2002[c]
Australia (includes ball clay)[c]	190	210	200	230
Austria (marketable)[c]	72	57	50	50
Belgium[c]	260[d]	300	300	300
Brazil (beneficiated)	746	1 067	1 517	1 820
Bulgaria	106	115[c]	98	100
Colombia (includes common clay)	1 984[c]	7 300	8 000[d]	8 500[c]
Czech Republic	XX[e]	2 800	5 183	5 500
Egypt	193	293	290	300
France (marketable)	344	345	325	300
Germany	684	1 925	1 800	1 800
Greece	189	69	60	60
India (processed and saleable crude)	741	713	670	710
Iran	150	266	837	800
Kazakhstan[c]	XX	XX	70	70
Korea, Republic of	1 755	2 792	1 858	2 381[d]
Malaysia	187	211	214	258[d]
Mexico	167	222	490	680
Nigeria[c]	1	105[d]	110[d]	110
Pakistan	45	31	65	50
Paraguay	74[c]	66[c]	67	67
Poland (washed)	44	53	89	142
Portugal[c]	150[d]	180	175	175
Russia (concentrate)	XX	50	41	45
Serbia and Montenegro (crude and washed)	XX	62	43[c]	70[c]
South Africa	134	147	122	91[d]
Spain (marketable, crude and washed)[f]	538[c]	316	320[c]	350[c]
Thailand (beneficiated)	256	461	113	165
Turkey	187	490	450	600
Ukraine	XX	950	222	225
United Kingdom (sales)[g]	2 911	2 586	2 303	2 400

Table 3 (Contd)

Country	Kaolin production (kilotonnes)			
	1991	1995	1999	2002[c]
USA[h]	9 570	9 480	9 160	8 010[d]
USSR[i]	9 000	XX	XX	XX
Uzbekistan[c]	XX	5 500	5 500	5 500
Vietnam[c]	1	1	398	600
Total	32 600	37 600	41 500	43 200

[a] From USGS (1997); Virta (1997, 2002).
[b] Figures are rounded to the nearest 1000 tonnes and may not add to totals shown. Table reports data available through August 21, 2003. Countries producing more than 430 000 tonnes (approximately 0.1% of the world total production) have been listed. In addition to the countries listed, China is known to have produced and Morocco and Suriname may also have produced kaolin, but information was considered inadequate to make reliable estimates of output levels (USGS, 1997; Virta, 1997, 2002).
[c] Figures are estimated, unless otherwise indicated (see footnote d).
[d] Reported figure.
[e] XX = Not applicable.
[f] Includes crude and washed kaolin and refractory clays not further described.
[g] Dry weight.
[h] Kaolin sold or used by producers.
[i] Dissolved in December 1991. This commodity is believed to be produced mainly in Ukraine and Uzbekistan.

Because of the varying composition of raw kaolin and different uses, raw kaolin generally requires processing (flotation, sedimentation, baking, etc.) to acquire characteristics suited to specific industrial uses.

3.2.1.3 Other clays

Illite is often an important or dominant component of common clay (Murray, 1994). Common clay is produced worldwide and is a major industrial mineral. In the USA, common clay and shale are produced in 41 states. In 2002, this production was 23 million tonnes and had a value of $148 million (Virta, 2002).

Illite occurs in commercially valuable purity and quantity in Malaysia (Bidor area of Perak), the USA (Illinois), the United Kingdom (South Wales), and Hungary (Tokaj Mountain in the north-eastern part of the country) (Grim, 1968; Fan & Aw, 1989). The most valuable ore is pure white and contains minimal iron. The commercial utility and thus the value of illite deposits may be reduced by lack of homogeneity in mineral content. There is no information on the annual production of relatively pure illite worldwide or by country.

3.2.2 Uses

3.2.2.1 Bentonite

Bentonite has many applications to a broad range of industrial and other activities. Usage of domestic production in the USA in 1995, 1999, and 2002 is summarized in Table 4. Major domestic uses, which include binding foundry sand (i.e., in moulds for castings), absorbing grease, oil, and animal wastes, pelletizing taconite iron ore, and improving the properties of many drilling muds, were 79% of the total. Most of the bentonite (79%) exported from the USA was used in foundry sand and drilling mud. Speciality uses of bentonite include serving as an ingredient for ceramics; waterproofing and sealing in civil engineering projects (e.g., blocking seepage loss from landfill sites, nuclear waste repositories, irrigation ditches, treatment ponds, and the like); serving as a filler, stabilizer, or extender in adhesives, paints, cosmetics, medicines, and other products, as a carrier in pesticides and fertilizers, and as a bonding agent in animal feeds; clarifying wine and vegetable oil; and purifying wastewater (Patterson & Murray, 1983; Kuzvart, 1984; Hosterman & Patterson, 1992; Hanchar et al., 2004). Small amounts of bentonite are also used as a catalyst in the refining of petroleum.

The use of sodium bentonite in hazardous waste containment utilizes a number of its special properties (Jepson, 1984). Its swelling ability makes it an effective soil sealant, since, by swelling within the interstices of the soil with which it is mixed, bentonite plugs the voids in the soil, creating a barrier of very low permeability. Swelling ability is aided by the possibility of a very small average particle size, allowing bentonite to plug even the smallest of voids. The high cation exchange capacity enhances the retention of wastes, especially heavy

Table 4. Usage of bentonite produced in the USA[a]

Use	Bentonite usage (kilotonnes)[b,c]		
	1995	1999	2002
Domestic			
Absorbents: Pet waste	574	1100	899
Absorbents: Other	88.4	W	W
Adhesives	W	2.17	2.1
Animal feed	97.8	49.9	42.4
Ceramics (except refractories)[d]	W	W	W
Drilling mud	627	733	762
Filler and extender applications[e]	69.9	47.1	45.7
Filtering, clarifying, decolorizing	W	91.6	127
Foundry sand	745	876	762
Pelletizing (iron ore)[f]	646	611	536
Miscellaneous refractories and kiln furniture	214	W	W
Miscellaneous[g]	94.9	81.8	117
Waterproofing and sealing	228	270	269
Total	3390	3870	3560
Exports			
Drilling mud	86.5	66.7	59.2
Foundry sand	256	238	244
Other[h]	89.1	124	106
Total	431	428	408
Grand total	3820	4290	3970

[a] From USGS (1997); Virta (1999, 2002).
[b] Data are rounded to no more than three significant figures and may not add to the totals shown.
[c] W = Withheld to avoid disclosing company proprietary data; included with "Miscellaneous."
[d] Includes catalysts and pottery.
[e] Includes asphalt tiles, cosmetics, ink, medical, miscellaneous filler and extender applications, paint, paperfilling, pesticides and related products, pharmaceuticals, and plastics.
[f] Excludes shipments to Canada.
[g] Includes waterproofing seals, chemical manufacturing, filtering and clarifying oils, heavy clay products, lightweight aggregate, water treatment and filtering, and other unknown uses.
[h] Includes absorbents, ceramics, fillers and extenders, filtering and clarifying oils, miscellaneous refractories, pelletizing refractories, waterproofing and sealing, and other unknown uses.

metals. In addition, a mixture of sodium bentonite and soil forms a tough, flexible mastic that is highly durable and not easily ruptured.

The ability of bentonite to bind cationic metals and certain pesticides has been applied experimentally to detoxifying victims of paraquat poisoning (Meredith & Vale, 1987) and to reducing the transfer of radiocaesium to milk and other animal-derived foods (Giese, 1989; Unsworth et al., 1989). Although effective in reducing the mortality of experimentally poisoned rats, there was no evidence that lavage with bentonite decreased human mortality following accidental ingestion of paraquat. Inclusion of bentonite in the diet of cattle reduced the transfer of radiocaesium into milk, but bentonite proved far less effective than cyanoferrates such as Prussian blue and thus was not the treatment of choice. In addition to these examples, there is an extensive literature documenting the ability of bentonite to adsorb these and other toxics. None of this research, however, appears to have led to major new uses for bentonite.

The chemical composition of bentonite affects its usage (Patterson & Murray, 1983; Kuzvart, 1984; Hosterman & Patterson, 1992). High-swelling bentonite, in which sodium is usually the dominant exchangeable ion, is preferred for drilling muds, pelletizing iron ore, and sealing and waterproofing, whereas low-swelling calcium bentonite is preferred for filtering, clarifying, and absorbing and for serving as a filler, stabilizer, extender, carrier, bonding agent, or catalyst. Both types are used as a foundry sand bond. Sodium bentonites provide good dry strength in moulds, whereas calcium bentonites provide good "green" (condition prior to drying) strength.

Bentonite is used in a large number of different cosmetic products, such as paste masks, skin care and cleansing preparations, eyeliners, foundations, and others. In 1998, bentonite was reported to be used in 78 different cosmetics in the USA, usually at concentrations between 1% and 10%, but reaching 80% in some paste masks (CIREP, 2003). Bentonite has been approved for use as a "Generally regarded as safe" (GRAS) food additive in the USA (US FDA, 2004).

The uses of montmorillonite-type Fuller's earth overlap those of bentonite (Hosterman & Patterson, 1992) (Table 5). Major uses include serving as an adsorbent for oil, grease, and animal waste and as a carrier for pesticides and fertilizers. Minor uses are filtering, clarifying, and decolorizing and serving as filler in paints, adhesives, and pharmaceuticals.

Table 5. Fuller's earth sold or used by producers in the USA, by use[a]

Use	Amount sold or used (kilotonnes)[b,c]		
	1995	1999	2002
Oil and grease absorbent	285	275	409
Pet waste absorbent	1530	1580	1580
Animal feed	72.7	82.9	81.5
Fertilizers	50.8	137	139
Fillers, extenders, binders[d]	75	63.9	58.8
Filtering, clarifying, decolorizing animal, mineral, vegetable oils, greases	9.07	W	63.3
Pesticides and related products	302	67.8	102
Miscellaneous[e]	130	245	264
Exports[f]	161	114	30.9
Total	2640	2560	2730

[a] From USGS (1997); Virta (1999, 2002).
[b] Data are rounded to no more than three significant digits and may not add to the totals shown.
[c] W = Withheld to avoid disclosing company proprietary data; included with "Miscellaneous."
[d] Includes adhesives, asphalt emulsions and tiles, gypsum products, medical applications, pharmaceuticals, cosmetics, paint, plastics, textiles, and other unknown uses.
[e] Includes catalysts (oil refining), electrical porcelain, drilling mud, roofing granules, chemical manufacturing, floor and wall tile, portland cement, refractories, and other unknown uses.
[f] Includes absorbents, drilling mud, fillers, extenders and binders, floor and wall tiles, mineral oils and greases, and other unknown uses.

3.2.2.2   Kaolin

In China, east of King-te-Chen city, kaolin was used for making porcelain in the third century BC.

In Europe, kaolin was used for making earthenware in about 5000 BC, but the method of making porcelain was unknown until about 1710. After that time, porcelain production spread first to Europe and then to the whole world (Kuzvart, 1984).

Like bentonite and common clay, kaolin is an important industrial mineral that has an enormous variety of uses. Uses of kaolin mined in the USA for the years 1995, 1999, and 2002 are summarized in Table 6. In all three years, about two-thirds of the total production was used domestically and the remainder exported. Use of kaolin as a coating for paper accounted for almost half of the total domestic consumption and for roughly 80% of the exported kaolin. Widespread use of kaolin-coated papers in the manufacture of cigarettes (Wynder & Hoffman, 1967) may expose smokers to kaolinite particles by inhalation. Other important uses of kaolin were as a filler in the production of paint, paper, and rubber, as a component of fibreglass and mineral wool, as a landfill liner, and as a catalyst in oil and gas refining. The usage historically associated with kaolin, manufacture of porcelain and chinaware, accounted for less than 1% of the domestic consumption in the USA. In China, 80–85% of the total production in 2003 was used for ceramics, 5% for paper, 3% for rubber, and 2% for paint. In India, 290 000 tonnes were used for ceramics, 84 000 tonnes for paints, 68 000 tonnes for paper/paperboard, 29 000 tonnes for detergents, and 27 500 tonnes for rubber (Moore, 2003).

Kaolinite has a number of properties relevant to medicine. It is an excellent adsorbent and will adsorb not only lipids and proteins (Wallace et al., 1975; Adamis & Timár, 1980) but also viruses and bacteria (Steel & Anderson, 1972; Schiffenbauer & Stotzky, 1982; Lipson & Stotzky, 1983). It can be used to induce aggregation of platelets (Cronberg & Caen, 1971), to initiate coagulation of plasma by activation of factor XII (Walsh, 1972), and to remove non-specific haemaglutinin inhibitors from serum (Haukenes & Aasen, 1972; Inouye & Kono, 1972). Kaolin is used in medical therapy as a local and gastrointestinal adsorbent (Kaopectate, bolus alba).

Kaolin is used in a large number of different cosmetic products, such as eyeshadows, blushers, face powders, "powders," mascaras, foundations, makeup bases, and others. In 1998, kaolin was reported to be used in 509 different cosmetics in the USA, usually at concentrations between 5% and 30%, but reaching 84% in some paste masks (CIREP, 2003). Medical, pharmaceutical, and cosmetic uses, however, accounted for roughly 0.01% of the total US consumption of kaolin (Table 6).

Table 6. Kaolin sold or used by producers in the USA, by use[a]

Use	Amount sold or used (kilotonnes)[b,c]		
	1995	1999	2002
Domestic			
Ceramics			
- Catalyst (oil and gas refining)	93.2	208	210
- Electrical porcelain	7.6	12.7	8.3
- Fine china and dinnerware	26.4	23.5	27.4
- Floor and wall tile	38.3	39.8	63.1
- Pottery	20.6	11.2	13.4
- Roofing granules	24.9	43.2	36.5
- Sanitary ware	67.9	75.6	85.2
- Miscellaneous	152	26.3	W
Chemical manufacture	130	23.2	31.6
Civil engineering	W	W	W
Fibreglass, mineral wool	402	329	288
Fillers, extenders, binders			
- Adhesive	71.6	81.5	67.4
- Fertilizer	W	W	3.55
- Medical, pharmaceutical, cosmetic	W	W	0.754
- Paint	270	288	298
- Paper coating	2800	3000	2540
- Paper filling	853	791	450
- Pesticide	11.2	13.1	W
- Plastic	39.5	39.7	49.7
- Rubber	194	222	177
- Miscellaneous	156	115	107
Heavy clay products			
- Brick, common and face	230	126	70.9
- Portland cement	W	54.2	W
Refractories			904[d]
- Firebrick, block and shapes	26.8	13.8	
- Grogs and calcines	190	135	

Table 6 (Contd)

Use	Amount sold or used (kilotonnes)[b,c]		
	1995	1999	2002
- High-alumina brick and specialties, kiln furniture	885	W	
- Foundry sand, mortar, cement, miscellaneous refractories	145	621	
Miscellaneous applications	138	430	91.6
Total	6970	6720	5520
Exports			
Ceramics	187	210	203
Paint	67.7	88.1	85
Paper coating	2040	1970	2040
Paper filling	145	110	93.9
Rubber	36.3	45.7	50.7
Miscellaneous[e]	165.9	23.8	19
Total	2510	2440	2490
Grand total	9480	9160	8010

[a] From USGS (1997); Virta (2001, 2002).
[b] Data are rounded to no more than three significant digits and may not add to the totals shown.
[c] W = Withheld to avoid disclosing company proprietary data; included with "Miscellaneous" or "Miscellaneous applications."
[d] Includes firebrick (blocks and shapes), grogs and calcines, high-alumina brick and specialties, kiln furniture, and miscellaneous refractories.
[e] Includes 145 000 tonnes of foundry sand, mortar, cement, and miscellaneous refractories in 1995.

3.2.2.3   Other clays

Common clay and shale, of which illite is often a major component, are used mainly in the manufacture of extruded and other bricks, portland and other cements, concrete blocks and structural concrete, and refractories. Highway surfacing, ceramic tiles, and ceramics and glass are other important uses (Virta, 2002; see Table 7). The relatively pure illite mined in Hungary, Malaysia, and the USA has a variety of uses. The illite produced in Malaysia is mainly exported to Japan, where it is mostly used to coat welding rods because of its fluxing properties. It is also used as a filler in sponge rubber and latex foam (Fan & Aw, 1989). Illite increases the plasticity and workability of the porcelain mass and produces a superior end product by filling the gaps in the kaolin crystals. Use in cosmetics requires a light-coloured illite with minimal iron. Fan & Aw (1988) suggested that illite could also serve as a less expensive substitute for feldspar in ceramics.

Table 7. Common clay and shale sold or used by producers in the USA, by use[a,b]

Use	Amount sold or used (kilotonnes)[c,d]		
	1995	1999	2002
Ceramics and glass[e]	139	181	174
Civil engineering and sealing	180	34.8	W
Floor and wall tile			
- Ceramic	301	400	387
- Other[f]	73	W	115
Heavy clay products			
- Brick, extruded	11 200	12 000	11 300
- Brick, other	1 640	1 800	1 500
- Drain tile and sewer pipe	136	27	39
- Flowerpots	48.6	W	46
- Flue linings	59.8	58.9	47.7
- Structural tile	21.5	22.7	W
- Other[g]	503	160	110
Lightweight aggregate			
- Concrete block	2 530	2 430	2 370
- Highway surfacing	248	317	364
- Structural concrete	869	929	908
- Miscellaneous[h]	521	259	361
Portland and other cements	6 400	5 010	3 950
Refractories[i]	459	785	795
Miscellaneous[j]	268	429	556
Total	25 600	24 800	23 000

[a] From USGS (1997); Virta (1999, 2002).
[b] Excludes Puerto Rico.
[c] Data are rounded to no more than three significant digits and may not add to the totals shown.
[d] W = Withheld to avoid disclosing company proprietary data; included with "Other" or "Miscellaneous."
[e] Includes pottery and roofing granules.

Table 7 (Contd)

f Includes quarry tile and miscellaneous floor and wall tiles.
g Includes flower pots, roofing tile, sewer pipe, structural tile, and miscellaneous clay products.
h Includes miscellaneous lightweight aggregates.
i Includes firebrick, blocks and shapes, grogs and calcines, mortar and cement, plugs, taps, wads, and miscellaneous refractories.
j Includes asphalt emulsions, exports, miscellaneous civil engineering and sealings, miscellaneous fillers, extenders and binders, pelletizing (iron ore), wallboard, and other unknown uses.

4. ENVIRONMENTAL LEVELS AND HUMAN EXPOSURE

4.1 General population exposure

4.1.1 Bentonite

Bentonite is widely distributed in natural materials that commonly occur as fine particles readily dispersed by air circulation and moving water. Bentonite is also widely used in an enormous variety of consumer products. Accordingly, general population exposure to low concentrations of bentonite must be widespread, if not universal. However, while Churg (1983) identified small amounts of kaolinite and illite in the lungs of a group of subjects drawn from the general population (see section 4.1.2), he did not detect montmorillonite. Similarly, while Dumortier et al. (1989) reported illite and kaolinite as ubiquitous in bronchoalveolar lavage (see section 4.1.2), they did not identify any particle as bentonite.

4.1.2 Kaolin and other clays

Kaolin and other clays are natural components of the soil and occur widely in ambient air as floating dust. Accordingly, exposure of the general population to them must be universal, albeit at low concentrations. In the vicinity of mines and industrial projects, kaolinite is likely to be present at high concentrations in air; no data are available, however.

The occurrence of mineral fibres in lung specimens from 20 persons who had no occupational dust exposure was determined on autopsy. Thirteen different minerals, among which were kaolinite and illite, were identified in the lungs of patients; 71% of the particles were fibres. There was no correlation between the number or type of fibres and age, sex, or smoking (Churg, 1983).

In a further study, Churg & Wiggs (1985) studied the relationship between mineral particles (and fibres) in the lungs (from autopsy) and lung cancer in men without occupational mineral exposure. The average number of mineral particles in the lungs of cancer patients was 525×10^6/g dry lung, and in the referents, 261×10^6/g dry lung. Kaolinite, talc, mica, feldspar, and crystalline quartz comprised the majority of particles in both groups. Approximately 90% of the particles were smaller than 2 µm, and 60% were smaller than 1 µm.

Dumortier et al. (1989) similarly reported kaolinite, illite, and mica in most of 51 subjects sampled via bronchoalveolar lavage. These subjects were all occupationally exposed to various industrial dusts, but not including kaolin or other clays to any great extent. The presence of illite in human lungs is also not a recent phenomenon, since it was found in the lungs of a 5300-year-old male human cadaver that had been well preserved by glacial ice (Pabst & Hofer, 1998).

Paoletti et al. (1987) identified 17 different mineral types in the lung parenchyma of 10 deceased subjects resident in an urban area and with no occupational dust exposure. Approximately 70% of the minerals consisted of phyllosilicates, in particular micas (10 cases), clays (kaolin and pyrophyllite, 10 cases), and chlorites (6 cases).

Samples of nine brands of tobacco and alveolar macrophages from cigarette smokers contained particles identified as kaolinite by energy-dispersive X-ray spectrometry (Brody & Craighead, 1975).

Kaolinite and mica particles were identified in the macrophages of the bronchoalveolar lavage fluid of 18 and 12, respectively, of the total of 22 patients for whom pulmonary lavage was performed. The frequency of these particles was higher among smokers and ex-smokers than among non-smokers (Johnson et al., 1986).

Mastin et al. (1986) determined the inorganic particles in the lungs of persons not occupationally exposed. Quartz, talc, and kaolinite could be detected in the lungs of practically all subjects (number of lungs examined: 48).

Kalliomäki et al. (1989) detected several different mineral types in the lungs from 11 unselected autopsy cases, including quartz, feldspar, kaolinite, and mica, reflecting the Finnish bedrock. Kaolinite particles constituted $10 \pm 5\%$ of the mineral particles in the lungs. More mineral particles were found in the lungs of non-smokers, but the differences were not statistically significant. Similarly, Yamada and co-workers (1997) also detected (fibrous) talc, mica, silica, chlorite, and kaolinite particles in the lungs of female lung cancer patients and referents from urban and rural areas in Japan.

4.2 Occupational exposure

Occupational exposure to clay dusts has been studied in many industries. However, numerical figures are usually given on total dust or respirable dust only, without analysis of the components of the dust. When information is given on individual components, it is mostly on silica. Exposure to silica in different trades has recently been extensively summarized (IARC, 1997b; IPCS, 2000) and will not be reviewed here. Clay particles have been detected in the lung tissue of bronchoalveolar lavage fluid of various worker groups (Churg, 1983; Churg & Wiggs, 1985; Johnson et al., 1986; Wagner et al., 1986; Dumortier et al., 1989; Dufresne et al., 1994; Gibbs & Pooley, 1994; Sébastien et al., 1994; Monsó et al., 1997). Exposure to clay dusts can also occur in small, family-run pottery shops — and may be considerably higher than in large workshops. Such exposure will not be considered in this review because of a lack of available information.

4.2.1 Bentonite

The information on occupational exposure to bentonite dust in mines, processing plants, and user industries (foundries) is summarized in Table 8. These data reveal a wide range of values for both total and respirable bentonite dust and indicate that very high total dust concentrations can be present in industrial settings involving both the production and utilization of bentonite. The highest values for total dust concentrations were reported by Melkonjan et al. (1981) in their study of bentonite mining and processing in Bulgaria. Only one of the eight types of locations they sampled, inside dump truck cabs during loading, had values consistently below 10 mg/m^3. Their highest reported values, 1150 and 1430 mg/m^3, and highest averages, 504 and 749 mg/m^3, obtained in the vicinity of packing and loading operations, reveal an extraordinary level of dustiness. These highest values are more than an order of magnitude greater than the highest values reported in any of the other studies summarized in Table 8. In the 1970s in foundries in Germany (Orth & Nisi, 1980), the dust concentrations in air were mostly below older observations for bentonite plants in the USA (McNally & Trostler, 1941; Phibbs et al., 1971), as well as far below contemporary observations from bentonite plants and

Table 8. Occupational exposure to bentonite dust

Location/date of observations	No. of observations / no. of facilities[a]	Dust concentration[b] (mg/m³)		No. of observations >10 mg/m³ total or >5 mg/m³ respirable	No. of observations of quartz >1%	Reference
		Range of observations	Average/range of facilities			
Fuller's earth plant, USA (Illinois) / 1934	5 / 1	1–19[c]	6	1	?[d]	McNally & Trostler (1941)
Bentonite plants, USA (Wyoming) / 1950–1967	17 / 4	1–92[c]	11–60	10	17	Phibbs et al. (1971)
Older iron foundry, Germany / pre-1980[e]	6 / 1	0.39–2.9 t	1.7	0	?	Orth & Nisi (1980)
	6 / 1	0.08–0.76 Rt	0.31	0	?	
	6 / 1	3.7–12.3 i	6.6	1	?	
	2 / 1	1.9–4.0 Ri	3.0	0	?	
Modern iron foundry, Germany / pre-1980[e]	5 / 1	0.56–2.8 t	1.6	0	?	Orth & Nisi (1980)
	5 / 1	0.15–0.43 Rt	0.23	0	?	
	4 / 1	1.7–7.9 i	4.8	0	?	
	2 / 1	0.81–3.2 Ri	2.0	0	?	
Bentonite mines and processing plants, Bulgaria / pre-1981	220 / ?	1.6–1430.0	333	Many	0	Melkonjan et al. (1981)

Table 8 (Contd)

Location/date of observations	No. of observations / no. of facilities[a]	Dust concentration[b] (mg/m³)		No. of observations >10 mg/m³ total or >5 mg/m³ respirable	No. of observations of quartz >1%	Reference
		Range of observations	Average/range of facilities			
Non-ferrous foundry, USA / pre-1987	13 / 1	0.77–27.9	6.65	3	At least 7	Que-Hee (1989)
		0.20–1.94 R	0.66	0	13	
Mines and processing plants, USA / 1992–1998	251 / 16	0.00–6.07 mainly R	0.06–2.18	1	85	US Mine Safety and Health Administration, personal communication (1999)
Mines and processing plants, Turkey, 1996–1999	24 / 11	2.6–11.2 R	?	At least 1	?	Turkish Ministry of Labor & Social Security, personal communication, 1999
Foundries, Turkey / 1990–1998	207 / 70	0.05–33.9 R	?	At least 8	?	communication, 1999

a Facilities may include mines, processing plants, and user industries.
b Values are for total dust unless indicated as respirable dust by R. i = impact sampler; t = continuous filter tape sampler using beta ray absorption; R = respirable dust.
c Original data in mppcf (million particles per cubic foot), converted by the approximate relationship 30 mppcf = 10 mg/m³.
d ? = no information.
e Authors believe that results obtained with the impact sampler (i) and total dust values from the tape sampler are flawed and underestimate the true values.

mines in Bulgaria (Melkonjan et al., 1981). In general, however, these data for total dust concentrations represent too limited a sampling to detect much in the way of regional, temporal, or other trends. Orth & Nisi (1980), however, concluded that their study of two German iron foundries demonstrated that change from the older, partially mechanized methods to modern, fully automatic, boxless moulding resulted in a reduction in dustiness. This may be true; however, in view of their problems in measurement methodology, the small differences between the two foundries in average dustiness, and the fact that only one foundry of each type was sampled, the conclusion seems premature.

Values for respirable bentonite dust are generally far below those for total bentonite dust (Table 8). However, values as high as 33.9 mg/m^3 have been reported in a foundry (Turkish Ministry of Labor & Social Security, personal communication, 1999). None of the observations from the two German foundries (Orth & Nisi, 1980) and only 1 of the 251 recent observations from the USA (US Mine Safety and Health Administration, personal communication, 1999) exceeded 5 mg/m^3, whereas the Turkish data for respirable dust in foundries suggest that this value is exceeded more frequently. Silica, which is widely viewed as the major hazard in many industrial dusts (Phibbs et al., 1971; Oudiz et al., 1983; Oxman et al., 1993), was a significant component of bentonite dust, especially in foundries.

4.2.2 Kaolin

Stobbe et al. (1986) analysed mine dusts of West Virginia, USA. Respirable dust samples collected in three locations in the mines contained 64% illite, 21% calcite, 8.5% kaolinite, and 6.7% quartz on average.

Processing of china clay involves considerable exposure (Tables 9 and 10). Exposure was especially significant before the 1960s (Sheers, 1964). Following drying, nearly all work stages were carried out in high dust concentrations (conveyor belts, bagging, storage in bulk). At present, ventilation is much more effective, and closed technologies are also widely used.

Lesser et al. (1978) analysed the free silica content of the airborne dust from a kaolin mill in Georgia and another from South Carolina and noted that while in most of the (total dust) samples free silica was

Table 9. Dust exposure in the kaolin industry

Time of sampling	Exposure group	No. and type of samples	Parameter measured	Result, average (mg/m^3)	Reference
1981	Miner	10; personal	Respirable dust	~0.2 (0.1–0.35)[a]	Sepulveda et al. (1983)
	Car loader	3; personal	Respirable dust	~1 (0.7–1.2)[a]	
	Car loader	3; area	Total dust	~10 (7–15)[a]	
	Bin operator	3; personal	Respirable dust	~2 (1.3–2.5)[a]	
	Bin operator	7; area	Total dust	~20 (10–30)[a]	
	Mill operator	3; personal	Respirable dust	~1 (0.8–1.2)[a]	
	Mill operator	8; area	Total dust	~20 (10–80)[a]	
	Baghouse labourer	5; personal	Respirable dust	~2 (0.3–5)[a]	
	Baghouse labourer	3; area	Total dust	~5 (1.5–9)[a]	
1951	Kaolin mining and processing		Respirable kaolin dust	Maximal value 377	Kennedy et al. (1983)
1960				Maximal value 361	
1979				All measurements <5	

Table 9 (Contd)

Time of sampling	Exposure group	No. and type of samples	Parameter measured	Result, average (mg/m³)	Reference
1977	Kaolin processing	14; personal	Respirable dust	3.9[b,c]	Altekruse et al. (1984)
1978		9; personal		3.8[b,c]	
1980		5; personal		5.3[b,c]	
1981		14; personal		1.8[b,c]	
1981		68; personal		1.74[b]	
1980	Maintenance area	15; personal		0.9[b,c]	
1981		11; personal		0.1[b,c]	
1981	Kaolin mine	12; personal		0.8[b,c]	
1981		5; personal		0.1[b,c]	
1980		4; personal		0.14[b]	
1990	"UK China Clay industry, potentially dusty occupations"	8000, both breathing zone and area sampling	Respirable dust	1.6	Comyns et al. (1994)
1978	Attritor mill	Altogether 500 personal samples annually for the industry	Respirable dust	4.7 (9.32)[d]	Rundle et al. (1993)
1990				2.1 (3.36)[d]	
1978	Dryers			3.5 (5.41)[d]	
1990				1.7 (2.78)[d]	

Table 9 (Contd)

Time of sampling	Exposure group	No. and type of samples	Parameter measured	Result, average (mg/m^3)	Reference
1978	Calciners			3.5 (3.93)[d]	Rundle et al. (1993)
1990				2.23 (3.25)[d]	
1978	Slurry plants			1.6[e]	
1990				1.2 (2.26)[d]	
1984– 1986	Attritor mill	114; personal	Respirable dust	2.7	Ogle et al. (1989)
	Dryers	681; personal		1.9	
	Calciners	63; personal		2.5	
	Slurry plants	69; personal		1.1	
	Tube presses	5; personal		0.5	

[a] Average (range), estimated from graphical presentation.
[b] Mean concentration.
[c] Estimated from a graph.
[d] Mean (9th decile).
[e] Mean, based on two samples.

Table 10. Dust exposure in brick, pipe, and tile industries

Time of sampling	Exposure group	No. and type of samples	Parameter measured	Dust concentration, mean (range)[a] (mg/m^3)	Quartz concentration, mean (range) (mg/m^3)[a]	Reference
1990–1991	Office, canteen	18; personal	Respirable dust	0.4 (0.3–1.3)	0.04 (0.02–0.13)	Love et al. (1999)
	Supervisors	336; personal		0.6 (0.3–1.6)	0.05 (0.02–0.18)	
	Quarries, moulders	301; personal		1.0 (0.4–1.6)	0.11 (0.04–0.28)	
	Brick cutters	10; personal		1.0 (0.3–1.4)	0.12 (0.04–0.21)	
	Mixed (fitters, fork truck drivers)	164; personal		1.2 (0.4–5.9)	0.07 (0.02–0.15)	
	Packers, kiln brick layers	202; personal		1.5 (0.3–2.3)	0.11 (0.02–0.21)	
	Forklift drivers in kilns or driers	46; personal		1.6 (0.3–2.4	0.13 (0.04–0.22)	
	Pan mill operators	83; personal		1.7 (0.4–5.8)	0.18 (0.04–0.75)	
	Labourers	172; personal		1.7 (0.3–4.8)	0.15 (0.03–0.38)	
	Sand users (blasters, applicators)	47; personal		2.3 (0.4–4.4)	0.23 (0.03–0.36)	
	Clean-up squad	25; personal		6.9 (1.9–7.8)	0.27 (0.05–0.33)	
	Kiln demolition	3; personal		10[b]	0.62	

Table 10 (Contd)

Time of sampling	Exposure group	No. and type of samples	Parameter measured	Dust concentration, mean (range)[a] (mg/m³)	Quartz concentration, mean (range) (mg/m³)[a]	Reference
1985[c]	Low dust exposure: offsetter, fireman, drier operator	35; personal	Respirable dust	1.02 SD 0.41	Silica 0.7% (mean)	Myers et al. (1989a)
	Medium dust exposure: setter, clamp packer, packer, preparation plant, hyster operator, rock face machine	29; personal		2.18 SD 1.42	Silica 3.9% (mean)	
	High dust exposure: drawer kiln cleaner, maintenance worker, downdraught kilns	22; personal		6.71 SD 4.80	Silica 3.2% (mean)	
1979–1980[c]	Hoisting winch, feeder and circulatory crusher	28	Respirable dust Total dust	4.3 (mean) 9.3 (0.7–50.4)	6.1% 18.4%	Wiecek et al. (1983)
	Press receipt of green brick	22	Respirable dust Total dust	1.9 (mean) 4.2 (0.8–12.3)	5.8% 10.8%	
	Backing off and setting the green brick in the furnace	28	Respirable dust Total dust	6.3 (mean) 19.0 (0.8–61.6)	3.7% 5.8%	

Table 10 (Contd)

Time of sampling	Exposure group	No. and type of samples	Parameter measured	Dust concentration, mean (range)[a] (mg/m³)	Quartz concentration, mean (range) (mg/m³)[a]	Reference
	Boring the ready-made product from the furnace	26	Respirable dust	33.9 (mean)	2.42%	Rajhans & Budlovsky (1972)
			Total dust	118.9 (7.3–418.0)	4.7%	
	Furnace and disposal	22	Respirable dust	41.8 (mean)	2.88%	
			Total dust	197.9 (44.0–636.0)	4.6%	
1970–1971	Brick plant; shale as raw material; dusty operations	9	Respirable dust	2.39 (mean)	15% free respirable silica	
		16		3 (mean)	20%	
		5		3 (mean)	20%	
		5		2.6 (mean)	19%	
	Brick plant; shale and clay as raw materials; dusty operations	6		2.24 (mean)	21%	
		6		4.26 (mean)	14.7%	
		5		1.5 (mean)	7.5%	
	Brick plant; clay as raw material; dusty operations	5		1.2 (mean)	12.5%	
		8		1.05 (mean)	10%	
ca. 1948[d]	Brick plant, different tasks	9	Respirable dust[e]	2.8 (0.3–5.7)		Keatinge & Potter (1949)
		19	Total dust	143 (3.5–250)		

Table 10 (Contd)

Time of sampling	Exposure group	No. and type of samples	Parameter measured	Dust concentration, mean (range) (mg/m³)[a]	Quartz concentration, mean (range) (mg/m³)[a]	Reference
1973	Sanitary ware manufacture	15	Respirable dust	0.42–0.54 (range)	28–39% quartz	Rees et al. (1992)
1974	Sanitary ware manufacture	24		0.02–0.69 (range)	4.19	
1986	Sanitary ware manufacture	43		0.8–4.2 (range)	10.60	
1989	Sanitary ware manufacture	9		0.38–2.1 (range)	8.28	
1974	Tile manufacture	24		0.35–8.41 (range)	9–29	
1989–1990	Wall tile and small bathroom fitting production		Respirable dust			
	- Slip production	4		2.8 (1.5–6.1)[f]	0.27 (mean)	
	- Tile dust production	3		1.2 (0.5–2.1)[f]	–	
	- Bathroom fittings	9		1.2 (0.2–6.6)[f]	0.17–0.4 (range)	
	- Tile making	9		1.4 (0.7–2.8)[f]	0.16–0.19 (range)	
	- Biscuit firing and sorting	5		0.9 (0.7–4.2)[f]	0.13 (mean)	
	- Tile glazing and glost sorting	6		1.9 (0.9–3.0)[f]	0.06 (mean)	
	- Bathroom fitting glazing	2		2.1 (0.9–3.3)[f]	0.07 (mean)	

52

Table 10 (Contd)

a Unless otherwise noted.;

b One site only.;

c Brick workers;

d Sampling time not given in the paper, assumed from the year of publication.

e Below 5.5 μm.;

f Median (range).

not detectable, in some specimens (three baggers and a bulk loader) quartz was present at approximately 1%. They also said that in all but one of the plants, some baggers and loaders were exposed to about 1% free silica (no numbers provided).

Significant exposure is expected also in paper, rubber, and plastic production. When producing insulator materials, quartzite and feldspar exposure may occur (Parkes, 1982). Sahle et al. (1990) carried out measurements of airborne dust in a Swedish paper mill. Total dust exposure at the plant was generally less than 3 mg/m³. The respirable fraction of the total dust varied from 15% to 70%. The inorganic dust was 36% of the total dust, and 10–15% of the total fibres were inorganic. The respirable dust at the plant contained kaolinite fibres, wollastonite, talc and other silicates, and cellulose fibres.

Dust conditions in power stations were studied in Poland by Wojtczak et al. (1989). In 14 power stations, average respirable dust and total dust concentrations ranged from 0.45 to 6.95 mg/m³ and from 1.55 to 85.0 mg/m³, respectively. The mean content of free crystalline quartz was less than 10%. At the workplace where ash dust was also encountered, the presence of respirable fibre at concentrations below 0.2 fibre/cm³ was measured. Alpha-quartz and mullite were found in all ash samples. Occasionally, kaolinite and orthoclase were also found. Only at 5 power stations out of 14 was the mean respirable dust concentration below 1 mg/m³.

In ceramic factories, workers are exposed to mixed dust. Analysis of the workplace respirable dust of five Hungarian ceramic factories demonstrated the presence of kaolinite (19–69%) and crystalline quartz (7–30%). X-ray amorphous material amounted to 7–76%, nearly half of which was amorphous silica (Adamis & Krass, 1991).

4.2.3 Other clays

There is some relevant information on exposure to illite dust in coal mining and in the manufacture of bricks, but none concerning the mining and processing of relatively pure deposits of illite. Wiecek et al. (1983), in their study of six brick manufacturing plants in Poland, reported illite in workplace floating dust prior to the firing of the bricks. In a sample of 27 dusts from European coal mines, illite was present in most and comprised as much as 55% of the fine material (particles <1.65 μm in mean geometric diameter) in one sample group (Bruch & Rosmanith, 1985). Illite comprised 11% of the total dust and

46% of the mineral dust (i.e., excluding coal dust) in a sample from a coal mine in Saar, Germany, and was also present in the lungs of most and possibly all of a sample of 13 Saar coal miners (Leiteritz et al., 1967). Stobbe et al. (1986) reported that in dust samples collected in nine locations in a bituminous coal mine in West Virginia, USA, illite represented 49–80% of the mineral dust and averaged 64%. In another study of a West Virginia coal mine, illite ranged from 3 to 89% of the mineral dust and averaged 52% (Grayson & Peng, 1986). In addition, observations in Dumortier et al. (1989) discussed above in section 4.1.2 indicated that even in industrial workers not specifically exposed to clay mineral dusts, illite can form a significant fraction of the particles retained by the lungs.

Data from the various studies summarized above suggest that workers in some occupations are frequently exposed to illite particles, which may comprise a large or major portion of certain industrial workplace dusts.

5. KINETICS AND METABOLISM IN LABORATORY ANIMALS AND HUMANS

Clays are not distinct chemicals, but are complex mixtures with usually a large component of specific clay minerals, such as montmorillonite, kaolinite, or illite. No information was available on the kinetics or metabolism of montmorillonite, kaolinite, or illite as they occur in most occupational settings. However, Snipes and co-workers (1983a, 1983b) and Bailey et al. (1982) studied the kinetics of fused aluminosilicate particles, prepared from montmorillonite by heating, in mice, rats, dogs, and humans after inhalation exposure. The method consisted of exchange by radiolabel (^{85}Sr, ^{88}Y, or ^{134}Cs) of cations normally present in montmorillonite clay, aerosolization, and heating at 1100 °C to yield spherical particles in which the radiolabel was incorporated in the lattice and very resistant to leaching. The polydisperse aerosol could then be segregated to different monodisperse fractions as needed and the specific fractions resuspended in an aqueous medium and administered to the experimental animals or human volunteers by inhalation.

The initial deposition of particles in the nasopharynx was lower in dogs than in mice or rats and increased with increasing activity median aerodynamic diameter (AMAD) (Table 11), whereas the tracheobronchial deposition was low and independent of animal species and particle size. The pulmonary deposition was remarkably higher in dogs than in mice and rats and decreased with increasing particle size; the change with size was especially marked in rats and mice.

The removal of particles from the lungs took place by solubilization *in situ* (leading to label distribution in the blood and carcass and excretion in the urine) and by physical clearance. For dogs, the overall clearance was slow, and the major mechanism of removal was solubilization. Rats and mice removed the label from the lungs rapidly, and the mechanism was mainly mechanical removal. The mechanical removal in rats and mice was mostly via the gastrointestinal tract, while in dogs it was mainly to the lung-associated lymph nodes. The mechanical clearance was independent of the particle size, while solubilization was faster for smaller particles. The initial half-time for the mechanical clearance was 140 days for dogs, which increased by 200 days post-exposure to 6900 days. In rats and mice, the initial half-time for mechanical clearance was 35 days and increased to 690 days in rats and 460 days in mice by approximately 500 days post-exposure.

Table 11. Deposition of [134]Cs-labelled aluminosilicate particles in dogs, rats, and mice after inhalation exposure[a]

Animal species	Particle size		Percentage of initial body burden deposited		
	AMAD[b] (µm)	Geometric diameter (µm)	Naso-pharynx	Tracheo-bronchial	Pulmonary
Dog	0.7	0.35	30	5	60
	1.5	0.90	40	5	50
	2.8	1.8	40	5	50
	Poly-disperse[c]	–	40	5	50
Rat	0.7	0.35	55	5	20
	1.5	0.90	62	5	13
	2.8	1.8	67	5	8
	Poly-disperse[c]	–	63	5	12
Mouse	0.7	0.35	45	5	30
	1.5	0.90	62	5	13
	2.8	1.8	67	5	8
	Poly-disperse[c]	–	62	5	13

[a] From Snipes et al. (1983a).
[b] Activity median aerodynamic diameter.
[c] AMAD and geometric standard deviation approximately 1.5–3 µm and 1.5–2.0, respectively.

Radiolabel from fused aluminosilicate particles, instilled in the apical lung lobe in dogs, was transferred mainly to the lateral tracheobronchial lymph nodes; after instillation to the cardiac or diaphragmatic lung lobes, radiolabel was mainly in the middle tracheobronchial lymph nodes (Snipes et al., 1983b). More recent studies indicate that ultrafine particles (<100 nm) have a high deposition in the nasal area of rats (Gerde et al., 1991) and humans (Cheng et al., 1988). Ultrafine particles can penetrate the alveolar/capillary barrier (Nemmar et al., 2004).

After inhalation exposure of seven volunteers to monodisperse fused aluminosilicate particles (1.9 or 6.1 μm aerodynamic diameter), there was a rapid initial clearance from the lung region of 8% and 40% of the two particle sizes, respectively, over 6 days, which was considered to represent deposition in the conducting airways. Of the remaining activity, 4% of the smaller and 11% of the larger particles cleared initially with a half-time of about 20 days, and the rest with half-times of 330 and 420 days. Taking into account the *in situ* dissolution (estimated to be 0.001 and 0.0005/day for the two particle sizes), the mechanical clearance half-time during the slow clearance stage is approximately 600 days and is independent of the particle size.

Two limited studies have addressed the leaching and bioavailability of metals from clays (Mascolo et al., 2004; Wiles et al., 2004). No significant differences were observed in the contents of aluminium, antimony, barium, bromine, caesium, calcium, cerium, chromium, cobalt, copper, dysprosium, europium, hafnium, iron, lanthanum, lutetium, magnesium, manganese, neodymium, nickel, samarium, scandium, selenium, sodium, strontium, sulfur, tantalum, tellurium, terbium, thorium, titanium, uranium, vanadium, ytterbium, zinc, or zirconium in the brain, kidney, liver, or tibia from pregnant SD rats dosed with 2% sodium montmorillonite or calcium montmorillonite clay compared with animals fed the basal diet. The main element components of the clays were aluminium (10%), iron (3%), and magnesium (0.5%) (as well as sodium in the sodium montmorillonite, 1%), with small amounts (usually less than 0.1%) of barium, caesium, manganese, strontium, zinc, and zirconium. The authors concluded that at this dietary level, the clays did not liberate significant amounts of trace elements (Wiles et al., 2004). Another study suggested that when rats were given 450 mg/kg of body weight of (three different varieties of) bentonite by gastric lavage for 6 days, contents of arsenic, cadmium, mercury, selenium, stibium, and tellurium in different organs were higher than in animals given normal diet (Mascolo et al., 2004). Because of limited reporting, the results are difficult to interpret.

6. EFFECTS ON LABORATORY MAMMALS AND *IN VITRO* TEST SYSTEMS

Clays are not distinct chemicals, but are complex mixtures with usually a large component of specific clay minerals, such as montmorillonite, kaolinite, or illite. They may also contain entities with well characterized adverse effects on health, such as quartz. In studies on the effects of clays or clay minerals on laboratory animals and even in *in vitro* tests, it is often not known what it was precisely that was studied and the impurities present in the material used. Thus, the results must be interpreted carefully. In this chapter, an effort is made to describe the material studied, but many of the studies, especially the older ones, give few details.

6.1 Single exposure

6.1.1 Bentonite

Data on single exposures of experimental animals to bentonite via intratracheal and intradermal administration, summarized in Table 12, indicate a variety of harmful effects. Bentonite has also been administered to animals by the intravenous and intramuscular routes (Boros & Warren, 1970, 1973), but these studies were not destined for, and cannot be used for, the assessment of health risks from bentonite.

Intratracheal injection of moderate amounts (0.16–0.27 g/kg of body weight) of finely powdered (where specified, mean diameter <5 μm) bentonite into rodents evoked responses indicative of cytotoxic effects, as well as rapid and/or long-lasting irritation of or damage to the lungs. Gross effects on the lungs of rats administered 0.19 g/kg of body weight included inflammation and oedema within 12 h, bronchopneumonia by day 1, and a doubling of wet weight by day 3, followed by a gradual decline to a value still significantly above the initial weight by day 90 (Tátrai et al., 1983). The histology of the inflammation included large numbers of neutrophil leukocytes plus some macrophages and lymphocytes at 12 h, mainly eosinophil and necrobiotic leukocytes by day 1, and collapse of the reticular network by day 3. Some bentonite particles were held within the macrophages and leukocytes at day 3.

Table 12. Effects of single exposure to bentonite (montmorillonite) in mammals

Species	Sex[a]	Dose (g/kg of body weight)[b] and dosing regimen / period of observation	Findings[c]	Reference
Intratracheal				
Rat (strain not identified)	M	Dose not specified,[d] diameter mainly <5 µm / 365 days	Three dusts with no or low concentrations of quartz and cristobalite gave rise only to storage foci; a fourth dust with the highest concentration of quartz (65%) produced productive fibrosis.	Timár et al. (1966)
Rat (strain not identified)	n.s.	40 mg[e] (0.16), diameter <5 µm / 365 days	Storage foci in lung after an unspecified period not greater than 365 days.	Timár et al. (1979)
Rat (strain not identified)	n.s.	60 mg[f] (0.24), particle size not specified / 6 months	At 6 months, storage foci in lung.	Timár et al. (1980)
Rat (CFY LATI-Gödöllő)	n.s.	60 mg[g] (0.24), particle size 0–2 µm; lung pathology studied after 3 or 10 months	At 3 and 10 months, sporadic storage foci of macrophage-derived cells; loose reticular network around the foci; no mitotic figures in the cells of the foci; no significant increase in mitotic figures in septal cells.	Ungváry et al. (1983)
Rat (CFY LATI-Gödöllő)	M	40 mg[h] (0.19), diameter <2 µm / 90 days	Lung weight nearly doubled at day 3; at day 90, decreased from weight at day 3, but still significantly above initial weight. **Histology:** Within 12 h, inflammation, perivascular oedema, and incipient bronchopneumonia; inflammatory reaction mainly neutrophil leukocytes plus some macrophages and lymphocytes. At day 1, broncho-pneumonia, necrobiotic leukocytes, and eosinophil leukocytes. Reticular network collapsed by day 3. Some	Tátrai et al. (1983)

Table 12 (Contd)

Species	Sex[a]	Dose (g/kg of body weight)[b] and dosing regimen / period of observation	Findings[c]	Reference
			bentonite particles held within macrophages and leuko-cytes at day 3. At day 90, bentonite in dust storage foci and foci surrounded by dense reticular network.	
Rat (CFY LATI-Gödöllö)	M	60 mg[h] (0.22), diameter <5 μm / 12 months	Acid phosphatase activity increased at day 3, but not at 1 month or 12 months; fibrosis not classifiable at day 3 and Belt-King grade 1 at 1 month and 12 months.	Tátrai et al. (1985)
Rat (Sprague-Dawley)	M	5–45 mg modified montmor-illonite[i] (0.017–0.15), median diameter 0.84 μm / 90 days	Through 90 days, no change in behaviour, body weight, or lung weight; at 90 days, gross lesions in lung; increased number of PAM; most instilled particles in alveoli and alveolar ducts and within PAM; mild fibrosis.	Schreider et al. (1985)
Rat (Sprague-Dawley CFY)	M	10 mg[j] (0.027), diameter <5 μm, median diameter 1–2 μm / 60 days	No effect on soluble protein or LDH in lavage from lung in samples taken on days 15, 30, and 60.	Adamis & Krass (1991)
Rat (SD)	M	10 mg of a kaolin containing 86% kaolinite, 4% quartz, 5% illite, and 5% amorphous silica	Bronchopulmonary lavage after 14, 30, and 90 days, number of cells, activities, LDH activity, protein, and phospholipid content of supernatant measured. Kaolinite caused a statistically significant increase in the protein content at 14 days, which returned to control level there-after. No changes were observed in the other parameters studied.	Adamis et al. (1998)

Table 12 (Contd)

Species	Sex[a]	Dose (g/kg of body weight)[b] and dosing regimen / period of observation	Findings[c]	Reference
Rat (Alderley Park Strain 1 SPF)	n.s.	0.5 mg[k] (0.0012), average diameter 3 μm / 7 days	In lavage from lung, large and rapid increase in PMN and LDH with maximum at day 2 and declining to somewhat above baseline at day 7; soluble protein was similar except for peak at day 1; AM showed increase by day 7.	Sykes et al. (1982)
	n.s.	5 mg[k] (0.012), average diameter 3 μm / 100 days	In lavage from lung, large and rapid increase in PMN to same level as with 0.5 mg dose, with peak at day 1 and a gradual decline to baseline by day 49; LDH and soluble protein also increased sharply, with a peak at day 1 and a subsequent decline, but had not returned to baseline value by day 100; AM increased sharply, with a peak at day 7, and declined slightly by day 100.	
Rat (Sprague-Dawley CFY)	M	60 mg[i] (0.27), diameter <7 μm, mainly 1–2 μm / 90 days	In lavage from lung, phospholipid and acid phosphatase activity elevated by day 3; phospholipid remained elevated at day 90 and acid phosphatase at day 30 (not reported after day 30).	Adamis et al. (1985)
Rat (Sprague-Dawley CFY)	M	60 mg[j] (0.22), diameter <2 μm / 12 months	At 3 months in lavage from lung, phospholipid and hydroxyproline somewhat elevated; relative abundance of phospholipid fractions unchanged.	Adamis et al. (1986)

Table 12 (Contd)

Species	Sex[a]	Dose (g/kg of body weight)[b] and dosing regimen / period of observation	Findings[c]	Reference
Mouse (strain not identified), 90–120 per group	n.s.	1, 10, or 100 µg bentonite[k] (5 × 10⁻³, 5 × 10⁻⁴, or 5 × 10⁻⁵), diameter 0.2–1.1 µm for 80% of the particles, median 0.4 µm injected intratracheally, followed by aerosolized group C *Streptococcus* sp.; mortality over 15 days recorded	Excess mortality 85% over controls from *Streptococcus* infection at instilled dose of 100 µg; 43% excess mortality with 10 µg; no significant change at 1 µg.	Hatch et al. (1985)
Intradermal				
Guinea-pig (out-bred albino)	n.s.	10 mg[k] (0.014), diameter 79–90 µm, injected into dorsal skin; some animals sensitized by injection of Freund adjuvant prior to injection of bentonite / 30 days	Sensitized and control animals showed identical responses; chronic inflammatory response with ulceration and maximum induration at 7–10 days. **Histology:** At day 2, vascular dilation and some infiltration of monocytes and PMN; by day 5, fairly dense macrophage infiltration between collagen bundles and some phagocytosis of bentonite particles; the lesion continued to evolve into a loose, disorganized mass of mainly multinuclear, phago-cytic cells, which by day 30 had phagocytized most of the bentonite; starting by day 10, active fibrosis ultimately walled off the lesion and subdivided infiltrating cells into clumps.	Browett et al. (1980)

Table 12 (Contd)

Species	Sex[a]	Dose (g/kg of body weight)[b] and dosing regimen / period of observation	Findings[c]	Reference
Rat (Wistar)	M	0.05 ml 5% bentonite gel[l] (0.01) injected under plantar surface of left paw / 42 days	Pronounced oedema in paw appeared within 7 h.	Marek (1981)
Rat (Wistar SPF)	M	0.05 ml 5% bentonite gel[m] (0.01) injected under plantar surface of left hind paw / 42 days	Pronounced oedema in paw within 1 h, continued to enlarge slowly for 1 week, then regressed slowly but was still evident at 6 weeks. **Histology:** Marked increase in PMN and macrophages peaking at about 30 h; PMN declines to baseline by 4 weeks; macrophages remained elevated through 6 weeks.	Marek & Blaha (1982)
Rat (Sprague-Dawley)	M	100 mg[n] (0.44), diameter <2 μm in 5-cm incision / 8 days	Incision failed to close; marked inflammation, development of foreign body granulomata, and collagen synthesis, compared with control incisions.	Cohen et al. (1976)
Guinea-pig[o]	n.s.	Dose not specified,[p] injected on both sides of back / 20 months	At autopsy (8 and 20 months), bentonite found encapsulated as a compact mass beneath skin.	Campbell & Gloyne (1942)
Guinea-pig	n.s.	5 mg[q] (0.01) in each of two incisions following a sub-infective dose (3200 cells) of *Staphylococcus aureus*	Infection established in 80–100% of montmorillonite-contaminated incisions (compared with about 0% in controls).	Rodeheaver et al. (1974)

Table 12 (Contd)

a M = male; n.s. = sex not specified.

b Value in parentheses is approximate dose in g/kg of body weight, estimated from the mean of the range of experimental animal weights (if given in the reference). Otherwise, an average weight of 250 g was assumed for rats, 20 g for mice, and 500 g for guinea-pigs.

c Abbreviations used: AM = alveolar macrophages; LDH = lactate dehydrogenase; PAM = pulmonary alveolar macrophages; PMN = polymorphonuclear leukocytes.

d Four mixed dusts were used; composition as weight per cent was estimated by microscopic examination: i) quartz 65%, montmorillonite 15%, kaolinite 10%, feldspar 10%; ii) quartz 30%, montmorillonite 55%, kaolinite 18%, feldspar 7%; iii) quartz 15%, cristobalite 30%, montmorillonite 45%, calcite 6%, feldspar 4%; iv) montmorillonite 98%, kaolinite 2%.

e Bentonite (76% montmorillonite, 3% quartz, 18% cristobalite, 3% feldspar).

f Bentonite (composition not given); vehicle for injection not given.

g Bentonite (montmorillonite 45%, quartz 15%, cristobalite 30%, calcite 6%, feldspar 4%) in 1 ml physiological sodium chloride containing 40 000 IU penicillin.

h Bentonite (73% montmorillonite, 18% cristobalite, 3% quartz, 3% feldspar) and 40 000 IU penicillin injected in saline.

i Montmorillonite (composition not given) treated to remove biologically active cations (Ca^{2+}, Mg^{2+}) and any organic matter; heated to 750 °C to collapse the lattice and inactivate any traces of silica; injected in physiological saline.

j Bentonite (73% montmorillonite, 18% cristobalite [incorporated in aluminium silicate], 3% crystalline quartz, 3% feldspar) injected in sterile isotonic saline.

k Bentonite (composition not given) injected in saline solution.

l Composition not given.

m Composition not given; rats received simultaneously an injection of 0.05 ml of 10% kaolin suspension in right paw.

n Sterile montmorillonite (composition not given).

o Two individuals.

p Bentonite Fuller's earth (85% montmorillonite, <1% quartz) in saline.

q Colloidal montmorillonite (particle size not specified) saturated with calcium, magnesium, or potassium (the specific cation had little effect on the results).

Changes in gross morphology and histology were accompanied by changes in the chemistry of lung lavage fluid. These included elevated levels of phospholipid (Adamis et al., 1985) and acid phosphatase (Adamis et al., 1985; Tátrai et al., 1985) by day 3. The increase in phospholipid was not, however, accompanied by any change in the relative abundance of phospholipid fractions (Adamis et al., 1986). Phospholipid was still somewhat elevated 3 months after exposure (Adamis et al., 1985, 1986). Tátrai et al. (1985) reported acid phosphatase back to baseline level at 1 and 12 months after exposure, whereas Adamis et al. (1985) found it still elevated at day 30. Hydroxyproline was slightly elevated 3 months after exposure (Adamis et al., 1986).

The effects of smaller intratracheal doses of bentonite on rats are uncertain. Adamis & Krass (1991) reported that 0.027 g/kg of body weight had no effect on protein or lactate dehydrogenase (LDH) in lavage from the lung 15–60 days after dosage, whereas Sykes et al. (1982), also using lavage from the lungs, noted a variety of effects indicating irritation and damage immediately following doses of 0.012 and 0.0012 g/kg of body weight. These effects included large and rapid increases in soluble protein, LDH, and polymorphonuclear leukocytes, with peaks at day 1 or day 2 followed by a decline to somewhat above baseline by day 7. Alveolar macrophages were increased by day 7. The findings in these two studies need not be contradictory, since the observations of Sykes et al. (1982) permit the possibility that elevated values for LDH and protein were present following dosage in the experiment of Adamis & Krass (1991), but declined back to baseline prior to their first sample on day 15.

The changes in gross morphology, white cell abundances, and biochemistry described above appear to form a generally coherent picture. Swelling and oedema stem from changes at the cellular level. Henderson et al. (1978a, 1978b) found increased LDH and protein in the lavage-sensitive indicators of lung damage that preceded visible mechanical damage and impairment of gas exchange. Increased LDH was ascribed to damage to cell membranes and leakage from lung cells, whereas increased protein was ascribed to increased vascular permeability. The latter conclusion is supported by analyses by Sykes et al. (1982), which reported serum albumin as the major constituent of the recovered protein. The increases in polymorphonuclear leukocytes may be triggered by phagocytosis of bentonite particles by alveolar macrophages (Terashima et al., 1997).

Single intratracheal exposures of rats to bentonite produced storage foci in the lungs 3–12 months later (Timár et al., 1979, 1980; Tátrai et al., 1983; Ungváry et al., 1983); even 12 months after exposure, fibrosis had not progressed beyond Belt-King grade 1 (Tátrai et al., 1985). Macroscopically, the foci were visible in fixed material as greyish-white masses up to the size of a "pin prick or a poppy seed" — i.e., approximately 1 mm (Timár et al., 1966). The structure of these storage foci was described in detail by Timár et al. (1966) using slides stained with haematoxylin-eosin, Van Gieson's solution, and Foot's silver impregnation. Timár et al. (1966: 5–6) reported that "The foci contained large cells with foamy protoplasm, and [with] nuclei often located at the edge of the cells. In the foam cells no lipids could be demonstrated by Sudan staining. PAS [*para*-aminosalicylic acid] reaction was intensely positive, showing that substances of carbohydrate nature are accumulated in the foci. In polarized light a great mass of intracellularly located doubly refractive fine grainy substance was to be seen. This characteristic alteration was to be observed in the lungs of the animals killed the 40th day after the application of the dust, and the microscopic findings in the tissues had not changed essentially up to the end of the experiment [365 days]. In the foci, sometimes a poorly developed reticular network of loose structure occurred too, but collagen fibres did not develop. The silver impregnated preparations showed in the area of the foci the precipitation of a greyish-black homogeneous substance between the fibres."

By ultrastructural analysis, Ungváry et al. (1983) augmented the above description, noting that whereas the cytoplasm of the peripheral cells of the foci was foamy, the closely packed cells of the centre with dislocated nuclei, which formed the bulk of the foci, had dark, granulated cytoplasm. No cell division was observed among cells of the foci 3 or 10 months after exposure. The cytoplasm of the dark, electron-dense cells contained a small number of mitochondria and occasionally other organelles plus a large amount of lamellar electron-dense material very similar in appearance to high phospholipid-containing breakdown products of intracellular origin. This material, however, by means of X-ray microanalysis, was demonstrated to be bentonite, which was swollen, presumably due to its montmorillonite content in direct contact with the cellular fluid. "The width of the electron dense and electronlucent layers depends on the degree of the swelling; the crystalline and the swollen ends of bentonite are spread out like cards" (Ungváry et al., 1983: 213). Cells of the foci with light foamy cytoplasm also contained bentonite, but in much smaller amounts. Both the above findings and the enzyme histochemical information on the foci (Ungváry et al., 1983) suggested that the metabolism of the foci is low. Ungváry et al. (1983) concluded that clearance of bentonite stored in foci from the lung is relatively slow and reflects the fact that cells of the foci are designed for long-term storage. The authors speculated that absence of toxic effects in the storage cells indicated that these cells have modified the surface characteristics of bentonite and coated the particles with protective molecules absorbed onto the bentonite.

Timár et al. (1966) provided evidence that the above cytotoxic responses leading to storage foci stem from montmorillonite and not from quartz or other impurities in the bentonite by comparing the effects of four mixed dusts containing varying proportions of montmorillonite, quartz, cristobalite, and other minerals. The composition of these dusts, essentially synthetic bentonites, is summarized in footnote d of Table 12. Dusts containing 55–98% montmorillonite and i) no quartz or cristobalite, ii) 30% quartz, or iii) 15% quartz and 30% cristobalite induced storage foci and no progressive fibrosis in rats following intratracheal administration. Only the dust with 15% montmorillonite and 65% quartz induced fibrosis. The authors ascribed the failure of the dust containing 30% quartz to produce productive fibrosis to a protective effect of the 55% montmorillonite, which they suggested may coat the surface of the quartz particles and thus inhibit the normal cytotoxic effects of the quartz. They further suggested a similar protective effect of montmorillonite against the cytotoxic effects of cristobalite. Havrankova & Skoda (1993) also reported that bentonite exerted a protective effect on rats against the toxicity of quartz dust. Observations by Schreider et al. (1985) supported the importance of molecular properties (crystalline structure and possibly exchangeable cations) in regulating the toxic effects of montmorillonite. They reported that an intratracheal dose (0.15 g/kg of body weight) of montmorillonite, washed to remove organic matter and biologically active cations and heated to 750 °C to collapse the crystal lattice and inactivate any traces of silica, induced not only increased pulmonary alveolar macrophages but also gross lesions in the lung and mild fibrosis at 90 days. This mild fibrosis, however, was not defined in terms of Belt-King grades.

In addition to directly damaging the lung, bentonite had a potential, even at low doses, for increasing susceptibility to pulmonary infection. Hatch et al. (1985) followed single intratracheal injections of 5, 0.5, or 0.05 mg/kg of body weight into mice with aerosolized Group C *Streptococcus* sp. (Table 12). The lowest dose did not significantly elevate mortality from *Streptococcus* infection over controls, whereas the highest dose produced 85% excess mortality and the intermediate dose 43% excess mortality over controls. Earlier studies by Hatch et al. (1981) demonstrated a close correlation between the effects on susceptibility to infection following dosing by sustained inhalation and by administering the same amount of test substance in a single intratracheal injection.

The mechanism by which bentonite increases susceptibility is unknown. Oberson et al. (1993), however, postulated that enzyme inhibition by bentonite may be a factor, based on their *in vitro* observation that a low concentration of bentonite can totally suppress the activity of human leukocyte elastase, an enzyme important in the destruction of microorganisms phagocytized by neutrophils.

In most experiments, bentonite also displayed local toxic effects after intra- or subdermal injection (Table 12). Subdermal injection of 10 mg/kg of body weight as 5% bentonite gel into the plantar of rats produced pronounced oedema within an hour (Marek, 1981; Marek & Blaha, 1982). The swelling continued to enlarge for a week and then slowly regressed, but it was still evident 6 weeks after injection. The oedema was accompanied by marked increases in macrophages and polymorphonuclear leukocytes, which peaked at 30 h. The leukocytes declined back to the baseline after 4 weeks, whereas the macrophages were still elevated at 6 weeks. Subdermally implanting 440 mg of sterile montmorillonite per kg of body weight into incisions in the backs of rats produced marked inflammation accompanied by the development of foreign body granulomata and collagen synthesis (Cohen et al., 1976). Incisions with montmorillonite failed to close during the 8-day duration of the experiment.

The granulomatous inflammation resulting from a moderate intradermal injection of relatively coarse (diameter 79–90 μm) bentonite into the skin of guinea-pigs was followed by Browett et al. (1980) both visually and with light and electron microscopy. The injection induced a chronic inflammatory response with ulceration and maximum induration at 7–10 days. By day 2, there was vascular dilation and some infiltration of monocytes and polymorphonuclear leukocytes. The macrophage infiltration rapidly increased in density and by the termination of observations, day 30, had become mainly multinuclear cells, had phagocytized most of the bentonite particles, and had evolved into a loose, disorganized mass. This mass was walled off from adjoining tissues by active fibrosis, which also subdivided the cells into clumps.

In a limited experiment involving two guinea-pigs and subdermal injection of bentonite Fuller's earth, the bentonite was found encapsulated as a compact mass beneath the skin at autopsy (Campbell & Gloyne, 1942). However, no adverse reaction to the injected bentonite was reported. McNally & Trostler (1941) also reported no adverse reactions following intraperitoneal injection of an unspecified dose of bentonite Fuller's earth into flank muscles of rabbits, other than adhesions in one animal.

As in the case of the lungs, montmorillonite also increased the susceptibility of subdermal tissue to infection (Rodeheaver et al., 1974). A dose of 0.01 g of sterile montmorillonite per kg of body weight in each of two experimental wounds in guinea-pigs 5 min after a subinfective dose of *Staphylococcus aureus* established infections in 80–100% of the incisions. Three separate experiments, which were run using montmorillonite saturated with calcium, magnesium, or potassium, yielded similar results. Almost none of the control incisions that received only *S. aureus* became infected.

6.1.2 Kaolin

6.1.2.1 Intratracheal administration

In the landmark paper in which intratracheal instillation as a means of studying the effects of dust on lungs was described, no collagen production was detected in the lung of a guinea-pig 336 days after the administration (whereas fibrosis was observed after similar injection of quartz) (Kettle & Hilton, 1932). Similarly, intratracheal instillation of kaolin from South Wales or untreated or ignited Cornish kaolin did not induce fibrosis in rats (King et al., 1948) (Table 13).

Table 13. Effects of intratracheal instillation of kaolin and illite on the respiratory tract

Species / gender[a] / number	Treatment	Findings	Reference
Guinea-pig / 2	Intratracheal instillation of an unstated amount of kaolin, follow-up for 14 and 336 days	No collagen production in 336 days; in animals treated similarly with quartz, fibrosis observed.	Kettle & Hilton (1932)
Rat / 12	Intratracheal instillation of a single dose of commercial acid-washed kaolin containing 8% hydrated free silica and 12% mica; two animals killed between days 3 and 6; the rest kept for life (up until 8 months)	Grade 2–3 fibrosis observed in rats after 8 months (grade 1 = minimal reticulin fibrosis, 4 = maximal fibrosis, as induced by quartz).	Belt & King (1945)
Rat / 6 or 10 per group	A single intratracheal instillation (50–60 mg) of washed South Wales kaolin, untreated Cornish kaolin, ignited Cornish kaolin, quartz; follow-up up to 6 months	South Wales kaolin (10 rats): Up to 60 days, no fibrous reaction; thereafter, local reticulinosis, no fibrosis or emphysema. Cornish kaolin (10): Eight animals died within 10 days; the remaining two showed no fibrous reaction. Ignited Cornish kaolin (6): Only four rats available for study; their survival was 14, 28, 73, and 140 days; in the last-mentioned, there was local reticulinosis. Quartz (6): Severe nodular silicosis in all five animals available for study; survival 68, 121, 130, 207, and 240 days.	King et al. (1948)

Table 13 (Contd)

Species / gender[a] / number	Treatment	Findings	Reference
Rat	Intratracheal administration of kaolin, kaolin baked for 1 h at 900 or 1200 °C	Active phagocytosis, local storage of the dust without reticular fibres, and nodules were observed. In the case of heat-treated kaolin samples, the reaction was somewhat stronger, although fibrosis reached grade 1 in only a few cases (Belt-King scale). Histological signs resembling silicosis did not develop.	Schmidt & Lüchtrath (1958)
Rat / 25 per group	A single intratracheal administration of kaolin containing kaolinite and quartz in ratios 82/18, 70/30, or 35/65; follow-up to 1 year	Foreign body reaction with the two samples containing 82 or 70% kaolinite, productive fibrosis with the sample containing 65% quartz.	Timár et al. (1966)
Rat / 10–15 per group	Single intratracheal instillation of kaolin (sericite and quartz as main impurities, 1% quartz), particle size <5 µm; histological analysis of lungs after 4 months	Cellular lesions, some loose reticulin, with either no collagen or a few collagen fibres.	Goldstein & Rendall (1969)
Rat / 10 per group	Intratracheal instillation of 50 mg non-specified kaolin, 3-month follow-up	Amount of collagen / amount of dust in lung after kaolin exposure 3 times higher than in titanium dioxide-treated control and 2.4 times more than in coal-treated animals, but only 15% of that in the quartz-treated animals.	Martin et al. (1977)

Table 13 (Contd)

Species / gender[a] / number	Treatment	Findings	Reference
Rat, strain not specified / number and gender not specified	40 mg of illite clay F, nominal composition 100% illite (diameter <2 or 2–5 µm), kaolin S (82% kaolinite, 18% quartz; diameter <3 µm), or kaolin Sz (95% kaolinite, 5% quartz; diameter <2 or 2–5 µm), instilled to rats; animals killed after 5, 15, 40, and 365 days and histological analysis performed	Illite F and kaolin Sz <2 µm caused "storage foci," kaolin S, "foreign body" reaction, and kaolin Sz 2–5 µm, mainly storage foci, rarely foreign body reaction.	Timár et al. (1979)
Rat, strain not specified / 10	60 mg of illite (not described) instilled intra-tracheally and followed for 6 months; lung weight, lipid, phospholipid, and hydroxyproline were analysed, and histological and histo-chemical studies on collagen performed	"Storage foci" observed in lungs of illite-treated rats.	Timár et al. (1980)
Rat, Sprague-Dawley / female / 10 per group	30 or 50[b] mg non-specified kaolin or illite clay injected intratracheally, followed for 3 and 12 months	Kaolin increased the lung weight 9 and 4 mg and collagen formed <26 and <7 mg/mg of injected dust at 3 and 12 months (all normalized to quartz = 100). Illite clay induced alveolar proteinosis and thus increased the lung weight 17 and 6 mg, and collagen formed <26 and 11 mg/mg of injected dust at 3 and 12 months (all normalized to quartz = 100).	Le Bouffant et al. (1980)

Table 13 (Contd)

Species / gender[a] / number	Treatment	Findings	Reference
Rat, Sprague-Dawley / female / 10 per group	30 or 50[b] mg non-specified kaolin or illite clay injected intratracheally, followed for 3 and 12 months	Correlation to haemolysis (rat erythrocytes) and release of LDH[c] and alkaline phosphatase from rabbit alveolar macrophages (see Table 17) weak.	Le Bouffant et al. (1980)
Rat, Fischer F344 / 10 per group	5 mg kaolin (non-specified), MMAD[d] 2.1 µm, instilled intratracheally, killed 1 day, 3 days, 7 days, 3 months, and 6 months later	Acute inflammatory reaction on day 1; thereafter, a slight interstitial cell thickening. At 3 and 6 months, lungs were normal.	Vallyathan et al. (1988)
Rat / male / 5 per group	Intratracheal injection of 10 mg of a kaolin containing 67% kaolinite and 23% quartz; bronchopulmonary lavage after 15 days, LDH activity and protein content and, after 15, 30, and 60 days, phospholipid content of supernatant measured	Kaolin did not induce significant LDH, protein, or phospholipid leakage to the supernate fraction. No change in the LDH or protein leakage was observed after similar exposure to eight other minerals, including two quartz samples. The quartz samples, however, increased the phospholipid content at 30 and 60 days.	Adamis & Krass (1991)
Rat, SD / male	Intratracheal injection of 10 mg of a kaolin containing 86% kaolinite, 4% quartz, 5% illite, and 5% amorphous silica; bronchopulmonary lavage after 14, 30, and 90 days, number of cells, activities, LDH activity, protein and phospholipid content of supernatant measured	Kaolinite caused a statistically significant increase in the protein content at 14 days, which returned to control level thereafter. No changes were observed in the other parameters studied.	Adamis et al. (1998)

Table 13 (Contd)

Species / gender[a] / number	Treatment	Findings	Reference
Mouse / 70 per group	Intratracheal instillation of 5 mg of kaolin, 91% >3.62 µm in diameter; follow-up 210 days	Fibroblast reaction from 60 days post-exposure, prominent from day 120. Grade 2 fibrosis by day 210 (Belt-King scale).	Sahu et al. (1978)
Rat, Wistar / female / 20 per group	Kaolinite 1 (K1) contained 2% muscovite; K2, 1% quartz and 9% muscovite; K3, <1% quartz and anatase and 1% muscovite; and K4, 1% quartz and anatase and 2% muscovite; mean value for volume distribution was 3.6 and 2.6 µm for K3 and K4, not analysed for K1 or K2; instilled once intratracheally at 50 mg/kg of body weight; autopsy after 7 months; lung weight, histology, amount of dust, hydroxy-proline, and total lipid content analysed; two samples of quartz also investigated, dose given 5 mg/kg of body weight	All kaolinite samples fibrogenic. Absolute amount of hydroxyproline roughly similar after exposure to kaolinites and quartz, hydroxy-proline / retained dust in kaolinite-treated animals 1/10 that in quartz-treated animals. Absolute increase of lung weight one-sixth and of total lipids roughly 10% of that in quartz-treated animals.	Rosmanith et al. (1989)
Rats, SD(SD)BR / male / 6 per group	Exposed by intratracheal instillation to 50 mg dust from the town Mexicali (see above), mean diameter 3.2 µm; lung analysis after 30 days	Multifocal interstitial lung disease. Mononuclear cell accumulation and presence of collagen fibres.	Osornio-Vargas et al. (1991)

[a] Where available.
[b] Composition or particle size not given.
[c] LDH = lactate dehydrogenase.
[d] MMAD = mass median aerodynamic diameter.

After the instillation of a single dose of commercial acid-washed kaolin containing 8% hydrated free silica and 12% mica, grade 2–3 fibrosis was observed in rats after 8 months (grade 1 = minimal reticulin fibrosis, grade 4 = maximal fibrosis, as induced by quartz) (Belt & King, 1945) (Table 13).

Following intratracheal dust treatment in rats, the histological reaction was found to depend on the composition of the dust (Tímár et al., 1966). Foreign body reaction as an effect of kaolinite was observed in all cases where the crystalline quartz content of the dust was less than or equal to 30%. The sample containing 65% quartz and 35% kaolinite caused progressive fibrosis. In addition to the composition, particle size also played a role in the development of the tissue reaction. Kaolin samples containing particles less than 2 μm caused storage foci (Tímár et al., 1979), while the kaolin samples containing bigger particles (particle size between 2 and 5 μm) caused mainly storage foci but also, to a smaller extent, foreign body reaction (Table 13).

Goldstein & Rendall (1969) observed cellular reaction with minimal fibrosis: some loose reticulin with either no collagen in some animals or a few collagen fibres in other rats 4 months after an intratracheal instillation of kaolin (Table 13).

Martin et al. (1977) observed an increase in the collagen content in the lungs of rats 3 months after an intratracheal instillation of non-specified kaolin; the reaction was considerably weaker than after a similar treatment with quartz (Table 13).

Sahu et al. (1978) described the development of grade 2 fibrosis in mice after 7 months of exposure to non-specified kaolin (Table 13).

Rosmanith et al. (1989) compared the fibrogenicity of four kaolinite samples with that of quartz in rats using intratracheal instillation. The fibrogenicity of kaolinites, as measured by the increase in hydroxyproline content in relation to the amount of dust retained, was approximately 1/10 that of quartz, but the inflammatory reaction was considerably less (Table 13).

No significant LDH, protein, or phospholipid leakage to the supernate fraction was observed in bronchoalveolar fluid 15–60 days after an intratracheal instillation of kaolin to rats. In this system, effects of quartz became apparent towards the end of the observation period (Adamis & Krass, 1991) (Table 13).

6.1.2.2 Parenteral administration

Policard & Collet (1954) demonstrated the development of reticulin fibres in rats 1–3 months after intraperitoneal administration of kaolin (90% <2 μm in diameter). The dust sample contained 1.2% free silica, an amount that the authors considered not to cause the effects observed.

Intraperitoneal administration of kaolinite (<3 μm or ~10 μm particle size; no quartz detectable with X-ray analysis) was fibrogenic in mice; the smaller particle size was just as active as quartz and led to fibrosis in 35 days, whereas fibrosis became evident only after 200 days for the larger size particles (Rüttner et al., 1952).

Intraperitoneally instilled kaolin and kaolin baked for 1 h at 900 or 1200 °C caused marked fibrosis. The degree of the reaction was only slightly smaller than that caused by quartz (Schmidt & Lüchtrath, 1958).

In a study of a large number of organic and inorganic particulates, kaolin (composition not specified, particle size 0.25–25 μm) injected intraperitoneally into rats produced a granulomatous reaction at 1 and 3 months, but no fibrosis. *In vitro*, it had low toxicity: it killed less than 2% of peritoneal macrophages and approximately 4% of alveolar macrophages (Styles & Wilson, 1973). For the group of some 20 particulate materials, the authors considered that there was a good correlation between toxicity to macrophages and fibrogenicity after intraperitoneal injection.

6.1.3 Other clays

Three and 12 months after an intratracheal instillation of illite clay with unknown quartz content, alveolar proteinosis and increased lung weight and collagen synthesis were observed in rat lungs (Table 13).

6.2 Repeated exposure

6.2.1 Bentonite

A case of profound hypochromic anaemia and mild hypokalaemia has been reported in a cat that habitually consumed large amounts of cat litter stated to contain 99% bentonite. The clinical status was dramatically improved by a transfusion and intravenous hydration. A similar episode was seen in the same cat a month later, when the use of the same litter material was resumed (Hornfeldt & Westfall, 1996). The causality is uncertain, as anaemia causes pica in cats (McDonough, 1997).

Bentonite has also been administered to animals using intravenous and intramuscular routes (Boros & Warren, 1970, 1973), but these studies were not destined for, and cannot be used for, the assessment of health risks from bentonite.

No long-term studies on the effects of bentonite in experimental animals using inhalation exposure are available.

In a study by Wilson (1953b, 1954), the reports of which provide only limited details, mice were fed bentonite (undefined) at 10, 25, and 50% of their diet (Table 14). The total numbers of mice in different treatment groups were not given, but of the 12 mice given 50% bentonite that survived 200 days or more, 11 developed tumours that the authors call "benign hepatomas" (without histological description). Eight of these mice had also been given thorotrast (known to be a liver carcinogen in mice, rats, and humans; IARC, 2001) before starting the diet study. Nine animals of the control group survived more than 200 days, and one developed hepatoma. In historical controls, no hepatomas were observed in 18 animals that had been given thorotrast and had lived >215 days; nor were hepatomas detected in 24 non-treated mice that had survived 484–575 days. Nine animals given low-protein, choline-deficient diet also survived more than 200 days, and three developed hepatomas. Development of hepatomas in these strains of mice on low-protein, choline-deficient diet had also been reported by the authors in an earlier study (Wilson, 1951).

Mice maintained on the diet with 50% bentonite also showed minimal growth and developed fatty livers and fibrosis of the liver

Table 14. Effects of repeated exposure to clays in experimental animals

Species / gender / number	Dose / study details / period of experiment	Findings	Reference
Bentonite			
Mouse, three inbred strains[a]:	Non-specified bentonite, at 10, 25, or 50% of the diet, and non-specified montmorillonite, at 50% of the diet, fed to mice for 60 days; body weight gain and liver histology studied	Fifty per cent bentonite and montmorillonite caused a 70–80% decrease in weight gain; 10% and 20% bentonite caused a smaller decrease in weight gain; at the high dose of bentonite, fatty liver developed.	Wilson (1953a)
2 females and 2 males per group in bentonite studies		Barium sulfate or cellulose at 50% of the diet did not lead to significant weight gain reduction; the effect was not due to smaller food intake.	
2 mice per group in montmorillonite and modified-diet studies		When choline or casein was added to the diet, the growth rate was partially restored.	
		When choline or casein was added to the diet containing 50% bentonite, the growth rate was partially restored and liver morphology remained normal.	

Table 14 (Contd)

Species / gender / number	Dose / study details / period of experiment	Findings	Reference
Mouse, three inbred strains / number and gender not indicated	Non-specified bentonite, at 10, 25, or 50% of the diet, fed to mice for an unspecified time (>200 days); body weight gain and liver histology studied; for liver imaging, some of the mice were given thorotrast (amount not given) prior to starting the dietary experiment	With 10% and 25%, slightly reduced growth; with 50%, minimal growth, fatty and fibrotic liver. At the highest dose, "benign hepatomas" in 9 of 12 mice that survived >200 days. Eight of the mice with tumours had also received thorotrast. Of the control mice, 9 survived >200 days, and 1 developed hepatoma.

Twenty-four historical controls that had survived 484–575 days were studied for hepatic tumours, and none was detected. Of historical controls with thorotrast injections, 18 had survived 216–511 days; no hepatomas were observed.

Of 9 mice given low-protein, choline-deficient diet that survived >200 days, 3 developed hepatoma. | Wilson (1953b, 1954) |
| **Kaolin** | | | |
| Guinea-pig / number and gender not given | Inhalation exposure to china clay "as supplied to the potteries" at unspecified concentration 2 h/day every day, every second day, twice weekly, or weekly; follow-up until 6 months | Until 3 months after the exposure, pulmonary reaction was slight, limited to mild alveolar proliferation. Thereafter, thickening of the alveolar walls, emphysema, and patchy bronchopneumonia with capillary congestion and massive eosinophil infiltration occurred. By 6 months, plaque formation and capillary bronchitis were observed. | Carleton (1924) |

Table 14 (Contd)

Species / gender / number	Dose / study details / period of experiment	Findings	Reference
Clay dust			
Rat, SD(SD)BR / male / 6 per group	Exposed by inhalation to dust from the town Mexicali, containing illite, calcium montmorillonite, kaolinite, alpha-quartz (23%), and small amounts of feldspars, calcite, volcanic glass; diameter approximately 10 μm, exposure 12 mg/m^3 for 3 h, tissue analysis after 0 h, 24 h, 8 days, 1 month	Accumulation of mononuclear cells and macrophages by 24 h in the airways and alveolar spaces; disappearance of particles and cellular changes by 8 days.	Osornio-Vargas et al. (1991)

[a] Inbred strains maintained in the author's laboratory for investigation of physiology and pathology of mouse liver.

(Table 14). Mice maintained on diets containing 10 or 25% bentonite displayed slightly reduced growth rates and no hepatic tumours (Wilson, 1953b, 1954).

In a separate, 60-day study (Wilson, 1953a), the addition of choline or casein to the 50% bentonite diet was reported to partially prevent the development of fatty liver and cirrhosis and permitted an improved but still subnormal growth rate. The author concluded that bentonite binds choline and other nutrients so that they cannot be absorbed into the body, and this deficiency then induces liver damage and hepatomas (Wilson, 1951, 1953a, 1953b, 1954). Because of the mixed exposure, limited numbers of animals, and limited reporting, these studies cannot be used to assess the carcinogenicity of bentonite.

6.2.2 Kaolin

Carleton (1924) studied the effects of kaolin (see Table 14) by inhalation in guinea-pigs. Until 3 months after the exposure, mild alveolar proliferation only was observed. Thereafter, patchy bronchopneumonia occurred, with massive eosinophil infiltration. By 6 months, plaque formation and capillary bronchitis were observed.

Wagner and co-workers (1987) studied the carcinogenicity of palygorskite and attapulgite in an inhalation study and used "coating grade" kaolin as a negative control: 20 male and 20 female Fischer rats were exposed by inhalation (10 mg/m^3, 91.4% of the particles <4.6 μm in diameter) for 6 h/day, 5 days/week, until they died. However, at 3, 6, and 12 months, four rats were killed from each group for ancillary studies (leaving only 28 animals for the carcinogenicity study). At the end of the study, full autopsy and microscopic analysis of the lungs, liver, kidney, spleen, and other organs were performed. In two kaolin-exposed rats, bronchoalveolar hyperplasia but no benign or malignant tumours of the lungs or pleura were observed. However, the number of tumours in the positive control group (crocidolite treatment at the same exposure level) was also low (one adenocarcinoma only). The mean fibrosis grading in the rats at interim sacrifices and at the end of the experiment was between 2.1 and 2.8 on a scale of 1–8 (1 being normal; 2, dust in macrophages; 3, early interstitial reaction; 4, first signs of fibrosis; 5, 6, 7, increasing fibrosis; 8, severe fibrosis).

As another part of the study on the carcinogenicity of mineral fibres, Wagner et al. (1987) injected a single dose (amount not given) of kaolin intrapleurally into 20 male and 20 female Fischer rats and followed the animals until moribund or dead (survival of animals in different groups not given). None of the kaolin-treated rats developed mesothelioma, whereas 34 of 40 of those given crocidolite did.

Mossman & Craighead (1982) treated cultured tracheas from hamsters with Georgia kaolin (composition not indicated; diameter 3–5 μm) and kaolin coated with 3-methylcholanthrene, implanted the tracheas after 4 weeks into syngeneic hamsters, and followed the animals until moribund at 105–110 weeks. Animals treated with kaolin did not develop tumours, but a high incidence of pulmonary tumours, often fatal, was observed in animals treated with kaolin coated with 3-methylcholanthrene. Animals treated with 3-methylcholanthrene-coated haematite or carbon particles also developed a similar spectrum of tumours (carcinomas, sarcomas, undifferentiated tumours).

6.3 Genotoxicity

No information was available on the genotoxicity of clays or clay minerals. Information on quartz has been summarized by IARC (1997) and IPCS (2001). Quartz did not test positively in standard bacterial mutagenesis assays. Results of genotoxicity studies of quartz conflict, and a direct genotoxic effect for quartz has not been confirmed or ruled out (IPCS, 2001).

6.4 Reproductive effects

In a study that provided limited details, Sprague-Dawley rats were administered calcium or sodium montmorillonite orally on pregnancy days 1–15. No effects were observed on the litter weight, implantation rates, or resorptions (Wiles et al., 2004) (Table 15).

Thirty-six Sprague-Dawley rats were fed a control diet, 20% kaolin (air-floated Georgia kaolin) diet, and iron-supplemented 20% kaolin diet 37–117 days before mating and during pregnancy (Patterson & Staszak, 1977). Dams receiving kaolin diet developed anaemia, whereas anaemia was not observed in dams receiving iron supplementation. The birth weight of the pups of the dams receiving kaolin was 9% smaller than that of the control dams ($P < 0.01$); again, iron supplementation prevented this decrease. There was no effect on the litter size or macroscopic malformations (Table 15).

Table 15. Developmental effects of bentonite (montmorillonite) in rats

Strain, number	Dose / period of experiment	Findings	Reference
Sprague-Dawley	2% calcium mont-morillonite or sodium montmorillonite in the diet + daily gavage with 1 ml of 0.25% aqueous clay sus-pension on preg-nancy days 1–15	No effect on maternal weight or maternal organ weights, litter weight, embryonic implantations, or resorptions.	Wiles et al. (2004)
Sprague-Dawley, 36	20% Georgia kaolin (crystal composition not given) in the diet 37–117 days before mating and during pregnancy	Maternal anaemia and reduction of pup birth weight observed; both reversed by iron supplementation. No terata observed, but details on the analysis not provided.	Patterson & Staszak (1977)

6.5 Administration with other agents

6.5.1 *Kaolin and microbes and microbe-derived factors*

In the study described in section 6.1.1 on bentonite (see Table 12), Hatch et al. (1985) also studied the effect of kaolin on infection resistance in mice. Intratracheal instillation of 100 µg kaolin (not further defined) caused a slightly (40%) and statistically not significantly increased mortality upon challenge by exposure to aerosolized group C streptococci by inhalation.

Intratracheal instillation of kaolin "prepared to the specification of the Society of Leather Trades" with atypical mycobacteria (not *Mycobacterium tuberculosis*) induced fibrotic reaction in the lungs of guinea-pigs (Byers & King, 1959). No control experiment with kaolin alone was performed.

Attygalle et al. (1954) studied the combined effects of dust samples characterized as mainly crystalline, to a small extent amorphous, aluminium silicate containing 77% kaolinite, 12% mica, and 8% quartz together with autoclaved *M. tuberculosis*. Kaolin samples alone did not cause macroscopic changes in rats or guinea-pigs.

Microscopically, a diffuse exudative reaction developed with giant cells containing dust particles after 30 days. The cells were prone to form groups. Noduli, however, were not found. Fine reticulin fibres were found in the cells. The lesions showed no progress and, after 500 days, were still of minimal grade 1. An avirulent strain (BCG) alone also produced a network of fine reticulin fibres, but no progressive fibrosis. In contrast, kaolin and BCG together produced, after 180–270 days, reactions of grade 2–3, with thickening of the alveolar walls and collagen fibre bundles. Thereafter, the lesions showed regression and, after 500 days, were of grade 1 only (reticulin fibre network with mononuclear cells).

Similarly, when kaolin (75 mg, not specified qualitatively) was injected into guinea-pigs intratracheally together with *Candida albicans*, thick reticulin fibres and collagenous fibrosis ensued; acute inflammatory reaction followed by formation of thick reticulin fibres were observed after kaolin alone (Zaidi et al., 1981).

6.5.2 *Kaolin and quartz*

Schmidt & Lüchtrath (1958) studied the effects of intratracheal administration of a 2:3 mixture of kaolin and quartz. Kaolin alone did not cause significant reaction in the lung. The mixture of kaolin and quartz caused significant changes in the lung; after 9 months, the fibrosis was grade 5. (Using pure quartz, the development of silicosis was somewhat quicker, but after 9 months, the difference compared with the kaolin–quartz mixture was insignificant.)

6.6 *In vitro* test systems

6.6.1 *Bentonite*

Experimental protocols in the studies summarized in Table 16 appear fairly uniform; cell suspensions were incubated with bentonite or montmorillonite for (where specified) 30 min to 24 h, and the cells were tested for physical integrity or functional ability. Cell types included rat peritoneal macrophages, rabbit and rat alveolar macrophages, a macrophage-like cell line, other white cells, human and other erythrocytes, rodent neural cells, hamster tracheal epithelial cells, and

Table 16. Effects of exposure to bentonite *in vitro*

System / species / gender	Dose (mg/ml)[a] / treatment	Findings[b]	References
Peritoneal macrophages			
Rat (Sprague-Dawley CFY)	Bentonite (73% montmorillonite, 3% quartz, 18% cristobalite; diameter not given) incubated with the cells for 24 h either as such or after dry milling for 32 h	Dry milling decreased the methylene blue adsorption to a third and the inhibition of TTC reduction by 75%.	Juhasz et al. (1978)
Rat (Sprague-Dawley CFY male)	Bentonite[c,d] (<5 μm diameter, median 1–2 μm) incubated in cell suspension for 24 h	More cytotoxic than quartz, other mineral dusts, or ceramic dusts as measured by decrease in TTC reduction activity.	Adamis & Timár (1976); Adamis & Krass (1991)
	Bentonite[c] 0.3 mg/ml (<5 μm diameter, median 1–2 μm) incubated in cell suspension for 3 h	Inert as measured by release of LDH into medium.	
	Bentonite[d,e] (<2 μm diameter) incubated in cell suspension for 3 h and 24 h	As cytotoxic as quartz in 3-h incubation and more cytotoxic than quartz in 24-h incubation as measured by decrease in TTC reduction activity.	Adamis & Timár (1978)
	Bentonite[e] 0.09, 0.27, 0.45 mg/ml (<2 μm diameter) incubated in cell suspension for 3 h	Intracellular and total (intracellular + medium) LDH activity decreased with increasing dose.	
	Bentonite[d,e] (<2 μm diameter) incubated in cell suspension for 24 h	More cytotoxic than quartz or other mineral dusts as measured by decrease in TTC reduction activity.	Adamis & Timár (1980)

Table 16 (Contd)

System / species / gender	Dose (mg/ml)[a] / treatment	Findings[b]	References
Rat (Sprague-Dawley CFY male)	Bentonite[d,e] (<2 μm diameter) incubated in cell suspension for 3, 4, and 24 h	After 3-h incubation with higher dose (not specified), intracellular LDH activity reduced, but no increase in LDH activity detected in medium; with concentration equal to one-third of higher dose, no reduction in intracellular LDH activity and no increase in LDH activity in medium after 4 h; after 24-h incubation with this lower dose, intracellular LDH activity reduced and LDH activity in medium increased.	Adamis & Timár (1980)
	Bentonite[d,f] with 3% and 34% quartz (<7 μm diameter, mode 1–2 μm) incubated[g] in cell suspension	Similar results with both bentonites; very cytotoxic as measured by decrease in TTC reduction activity; no loss of intracellular LDH to the medium detected.	Adamis et al. (1985)
	Bentonite[c,d] (mode diameter 1–2 μm) incubated[g] in cell suspension	Very cytotoxic as measured by decrease in TTC reduction activity; no loss of intracellular LDH to the medium detected.	
Alveolar macrophages			
Rabbit (New Zealand)	Bentonite[e] 1.0 mg/ml (0.4 μm median diameter) incubated in cell suspension for 20 h	Viability index (trypan blue exclusion) lowered by 65% ($P < 0.05$) and cellular ATP (luciferin–luciferase reaction) by 92% ($P < 0.05$).	Hatch et al. (1985)

Table 16 (Contd)

System / species / gender	Dose (mg/ml)[a] / treatment	Findings[b]	References
Rabbit (New Zealand)	Montmorillonite[e] 0.5 mg/ml (<5 µm diameter) incubated in cell suspension for 24 h	Complete (100%) haemolysis.	Daniel & Le Bouffant (1980)
Rat (strain not specified)	Bentonite[h] 1.0 mg/ml (<5 µm diameter, mode 2.7 µm) incubated in cell suspension for 2 h	Small increase above control in β-glucuronidase detected in medium; LDH and β-*N*-acetylglucosaminidase detected in medium less than control.	Vallyathan et al. (1988)
Neutrophils			
Human (multiple donors)	Montmorillonite[i] 10 mg/ml incubated[g] in cell suspension	Rapid and complete lysis, with most of LDH released from cells adsorbed onto the montmorillonite; pretreatment of montmorillonite with 1% human albumin for 30 min blocked adsorption without affecting lysis; pretreatment with 10% human albumin blocked both adsorption and lysis.	Dougherty et al. (1985)
Human phagocytic cells from one donor	Montmorillonite STx-1, 1% quartz and 10% cristobalite, particle size 2.5 µm median volume diameter	Kaolinite at concentrations in the order of 1 mg/ml induced luminol-dependent chemiluminescence as an expression of generation of reactive oxygen species in both monocytes and neutrophils when opsonized or when not opsonized.	Gormley et al. (1985)

Table 16 (Contd)

System / species / gender	Dose (mg/ml)[a] / treatment	Findings[b]	References
Leukocytes			
Human	Sterile montmorillonite[e] saturated with magnesium 5.3 mg/ml (<2 µm diameter) incubated in cell suspension for 30 min, then suspension inoculated with *Staphylococcus aureus* or *Pseudomonas aeruginosa*	Leukocytes lost all capacity to phagocytize and kill *S. aureus* and *P. aeruginosa*.	Haury et al. (1977)
Erythrocytes			
Rat (strain not specified)	Montmorillonite[e] 1.0 mg/ml (<5 µm diameter) incubated[g] in cell suspension for 1 h	Complete (100%) haemolysis.	Daniel & Le Bouffant (1980)
Sheep (citrated blood)	Bentonite[e] (median diameter 5 µm) incubated in cell suspension	Speed of haemolysis dependent on concentration of bentonite: 100% within 5 min at 0.2 mg/ml; approximately 12% within 30 min at 0.01 mg/ml.	Ottery & Gormley (1978)
Sheep	Bentonite[h] at four concentrations,[i] 0.1–0.7 mg/ml (<5 µm diameter, mode 2.7 µm), incubated in cell suspension for 50 min	Most haemolytic of mineral dusts tested; linear, dose-dependent response; about 7% haemolysis with 0.1 mg/ml; 84% with 0.74 mg/ml.	Vallyathan et al. (1988)

Table 16 (Contd)

System / species / gender	Dose (mg/ml)[a] / treatment	Findings[b]	References
Sheep (citrated blood, single donor)	Montmorillonite[k] at six concentrations,[l] 0.005–0.25 mg/ml (70% <1 μm diameter), incubated in cell suspension for 1 h	More haemolytic than silica or chrysotile; clear dose–response relationship; 5% haemolysis with 0.005 mg/ml; 95% with 0.25 mg/ml.	Woodworth et al. (1982)
Cow (erythrocytes washed with buffered saline)	Montmorillonite[e] at several concentrations (<50 μm diameter) incubated in cell suspension for 1 h	A concentration of 0.006 mg/ml induced 50% haemolysis of erythrocytes; particles 0.2–2 μm in diameter were most active; particles >2 μm diameter had little or no haemolytic activity; reduction of surface charge and cation exchange capacity by coating particles with an aluminium-hydroxy polymer largely eliminated haemolytic capacity.	Oscarson et al. (1986)
Human (citrated blood)	Bentonite[c] 1.0 mg/ml (<5 μm diameter, median 1–2 μm) incubated in cell suspension for 24 h	Almost complete (95%) haemolysis.	Adamis & Krass (1991)

Table 16 (Contd)

System / species / gender	Dose (mg/ml)[a] / treatment	Findings[b]	References
Human (multiple donors)	Montmorillonite[i] 0.6–20 mg/ml incubated[g] in cell suspension	Haemolysis 30% from incubation with an unspecified concentration of montmorillonite; 20% haemolysis using montmorillonite pretreated with 1% human albumin; pre-treatment of montmorillonite with 10% human albumin virtually blocked haemolysis.	Dougherty et al. (1985)
Human	Bentonites,[m] calcium montmorillonite,[e] montmorillonite[e] + quartz (<5 µm diameter) incubated in cell suspension for 1 h	Calculated concentration causing 50% haemolysis = 1.0 mg/ml for bentonite (a),[m] 1.66 mg/ml for bentonite (b),[m] 5.0 mg/ml for calcium montmorillonite, and 0.8 mg/ml for montmorillonite + quartz; haemolytic activity proportionate to surface area of mineral powder; haemolytic activity of minerals largely lost after heating to over 500 °C.	Mányai et al. (1969)
Human	Bentonites,[m] calcium montmorillonite,[e] montmorillonite[e] + quartz (<5 µm diameter) incubated in cell suspension for 1 h	Maximum haemolysis between pH 7 and pH 8, and much diminished below pH 7.	Mányai et al. (1969)

Table 16 (Contd)

System / species / gender	Dose (mg/ml)[a] / treatment	Findings[b]	References
Neural cells			
Neuroblastoma (N1E-115) cells with differentiation induced by dimethylsulfoxide	Montmorillonite[i] 0.1–1.0 mg/ml incubated briefly in cell suspension	Within minutes, resting potential depolarized and ability to maintain action potentials in response to stimulation was lost; within 30 min, severe morphological deterioration of cells.	Banin & Meiri (1990)
Neuronal (primary cells from 11- to 13-day-old mouse embryos)	Montmorillonite[i] and bentonite[i] 0.1 mg/ml incubated in cell suspension for 1 h	Clay particles rapidly collected on plasma membrane; after 1 h, cells largely lysed by montmorillonite and completely lysed by bentonite.	Murphy et al. (1993a)
Neuroblastoma (N1E-115 cells)	Montmorillonite[i] and bentonite[i] 0.1 mg/ml incubated in cell suspension for 18 h	Clay particles rapidly collected on plasma membrane; after 18 h, cells unharmed by montmorillonite or by bentonite.	Murphy et al. (1993a)
Neuroblastoma (N1E-115) cells and oligo-dendroglial (ROC-1) cells	Montmorillonite[e] and bentonite[e] 0.01, 0.03, 0.1 mg/ml (mainly 1–2 μm diameter) incubated in cell suspension for 1, 6, or 24 h	No lysis of cells or alteration of LDH activity in medium; slight release of fatty acids (mainly saturated) from cells; montmorillonite significantly decreased viability (assessed by trypan blue exclusion) of N1E-115 cells after 24 h, whereas bentonite had no effect.	Murphy et al. (1993b)

Table 16 (Contd)

System / species / gender	Dose (mg/ml)[a] / treatment	Findings[b]	References
Other cell types			
Tracheal epithelial (cloned cell line from Syrian hamster, strain 87.20) in log growth phase	Montmorillonite[k] at four concentrations, 0.003–0.1 mg/ml[n] (70% <1 µm diameter), incubated in cell suspension for 24 h	Cells phagocytized clay particles; dose-dependent damage to plasma membrane as evidenced by loss of ^{51}Cr from cells; loss of ^{51}Cr after 24 h ranged from approximately 12% with 0.003 mg/ml to 58% with 0.1 mg/ml.	Woodworth et al. (1982)
Human umbilical vein endothelial cells	Montmorillonite[e] and bentonite[e] 0.01, 0.03, 0.1 mg/ml (mainly 1–2 µm diameter) incubated in cell suspension for 1, 6, or 24 h	Both montmorillonite and bentonite associated with plasma membrane within 1 h; after 24 h, both caused lysis of cells, a large dose-dependent release of fatty acids (mainly saturated) from cells, and a significant increase in LDH activity in the medium.	Murphy et al. (1993b)
Macrophage-like cell line P338D$_1$	Three montmorillonites (Saz-1, Swy-1, STx-1) from Source Clays repository, particle sizes 2.5–3.1 µm; Saz-1 with no cristobalite or quartz, Swy-1 with 5% quartz, and STx-1 with 10% cristobalite and 1% quartz	Cell viability at 20 µg/ml 79% for Saz-1, 47% for Swy-1, and 21% for STx-1.	Gormley & Addison (1983)

93

Table 16 (Contd)

System / species / gender	Dose (mg/ml)[a] / treatment	Findings[b]	References
Artificial organelles			
Liposomes (artificial phospholipid membrane vesicles[o] 0.1–2 μm diameter) entrapping dissolved chromate (CrO_4^{2-})	Montmorillonite[k] at five concentrations,[p] 0.1–10 mg/ml (70% <1 μm diameter), incubated in cell suspension for 1 h	Dose-dependent loss of chromate from vesicles; loss of chromate after 1 h (in excess of spontaneous rate) ranged from about 6% with 0.1 mg/ml to about 44% with 10 mg/ml; spontaneous rate was 4–6%.	Woodworth et al. (1982)

[a] Dose recalculated as necessary to mg/ml.
[b] ATP = adenosine triphosphate; LDH = lactate dehydrogenase; TTC = 2,3,5-triphenyltetrazolium chloride.
[c] 73% montmorillonite, 18% cristobalite (incorporated into montmorillonite), 3% quartz, 3% feldspar.
[d] Dose not specified.
[e] Composition not specified.
[f] Composition not given except for quartz.
[g] Time not specified.
[h] Bentonite described as 90% pure with 0.8% silicon.
[i] Composition and particle size not specified.
[j] 1.0, 2.4, 5.0, and 7.4 mg/ml.
[k] >99% pure; <1% feldspar, biotite, and gypsum.
[l] 0.005, 0.010, 0.025, 0.05, 0.1, and 0.25 mg/ml.
[m] One bentonite (a) described as predominantly montmorillonite with small amounts of kaolinite and quartz; the other (b) as montmorillonite 25%, illite 50%, quartz 25%.
[n] 0.003, 0.01, 0.03, and 0.1 mg/ml.
[o] Prepared from dipalmitoyl phosphatidylcholine, sphingomyelin, cholesterol, and dicetylphosphate.
[p] 0.1, 0.3, 1, 3, and 10 mg/ml.

human umbilical vein endothelial cells. The results usually indicated a high degree of cytotoxicity. Concentrations below 1.0 mg/ml of bentonite and montmorillonite particles less than 5 μm in diameter caused lysis of human neutrophils, many types of erythrocytes, mouse embryo neuronal cells, and human umbilical vein endothelial cells. The velocity and degree of lysis of sheep erythrocytes were dose dependent. Using cow erythrocytes and montmorillonite, Oscarson et al. (1986) found that particles 0.2–2.0 μm in diameter were most active and that particles greater than 2.0 μm in diameter showed little or no haemolytic activity. The ability of low concentrations of these particles to lyse mammalian cells, although widespread, was not universal, since Murphy et al. (1993a, 1993b) reported no lysis of neuroblastoma cells or oligodendroglial cells incubated in 0.1 mg montmorillonite or bentonite/ml for 18 or 24 h.

Lytic activity appeared to stem from surface properties of the particles, since it was blocked by heating montmorillonite and bentonite to over 500 °C, which altered surface characteristics (Mányai et al., 1969), or by pretreatment of montmorillonite with 10% albumin, which coated the particles and thus altered surface characteristics (Dougherty et al., 1985). Mányai et al. (1969) also reported that the haemolytic activity of bentonite and montmorillonite was proportional to the surface areas of the mineral powder.

Other indications of damage to the physical integrity of cells from montmorillonite or bentonite particles were a dose-dependent loss of ^{51}Cr from hamster tracheal epithelial cells (Woodworth et al., 1982), a dose-dependent loss of fatty acids from human umbilical vein endothelial cells (Murphy et al., 1993b), and elevated loss of β-glucuronidase from rat alveolar macrophages (Vallyathan et al., 1988). Montmorillonite also caused a dose-dependent loss of chromate from artificial liposomes (Woodworth et al., 1982) and suppressed the activity of the enzyme human elastase.

Loss of LDH from cells is a sensitive measure of damage to cell membranes. Several authors, however, reported less loss of LDH to the supernatant in cell cultures incubated with bentonite or montmorillonite than in the controls (Table 16). Dougherty et al. (1985) explained this paradox by demonstrating that although LDH was greatly diminished in quantity in cells incubated with montmorillonite, LDH lost from cells was largely adsorbed on the montmorillonite particles and thus did not appear in the supernatant. Pretreatment of the montmorillonite with 1% human albumin blocked this adsorption without affecting lysis. These observations may also explain the low value of another enzyme used to measure cell membrane damage, β-N-acetylglucosaminidase, in the supernatant from suspensions of rat alveolar macrophages incubated with bentonite (Vallyathan et al., 1988).

In addition to the lytic and membrane damage effects described above, bentonite and montmorillonite also usually exerted other cytotoxic effects. Oberson et al. (1993) noted that montmorillonite particles totally suppressed hydrolysis of elastin by human leukocyte elastase. Adamis and his co-workers (Adamis & Timár, 1976, 1978, 1980; Adamis et al., 1985; Adamis & Krass, 1991) found that bentonite particles were highly cytotoxic to rat peritoneal macrophages as measured by a decrease in 2,3,5-triphenyltetrazolium chloride (TTC) reduction activity (Table 16). Hatch et al. (1985) reported that incubation of rabbit alveolar macrophages with bentonite greatly lowered the viability of the cells as well as cellular ATP. Murphy et al. (1993b) similarly found that montmorillonite significantly decreased the viability of neuroblastoma cells (assessed by trypan blue exclusion), but surprisingly that bentonite had no effect on these cells as measured by this test. The particle size of both minerals, mainly 1–2 μm diameter, was appropriate to produce adverse effects; the composition of the minerals, however, was not specified.

Functional effects of montmorillonite particles on mammalian cells were also reported. Neuroblastoma cells (with differentiation induced by exposure to dimethylsulfoxide) briefly exposed to up to 1.0 mg/ml had their resting potentials depolarized and lost the ability to maintain action potentials in response to stimulation (Banin & Meiri, 1990). Human leukocytes incubated with sterile, magnesium-saturated montmorillonite at a concentration of 5.3 mg/ml lost all capacity to phagocytize *Staphylococcus aureus* and *Pseudomonas aeruginosa* (Haury et al., 1977). Exposure of hamster tracheal epithelial cells to montmorillonite, however, did not prevent these cells from phagocytizing the clay particles (Woodworth et al., 1982).

6.6.2   Kaolin

6.6.2.1   Haemolysis

Different kaolin samples were found to have a significant haemolysing effect on red cells (Table 17). Mányai et al. (1970) investigated a water-washed Hungarian kaolin sample. The non-heat-treated samples had a haemolytic effect. The sample pretreated by heating at 200 or 350 °C for 90 min had a somewhat stronger haemolytic effect on the red cells than the original sample.

The heat treatment caused the loss of water adsorbed to the surface. Samples pretreated by heating at 500 or 650 °C lost the structurally bound water, and the haemolytic activity was practically abolished. Undoubtedly, the presence of hydroxyl groups is important with respect to the interaction between the red cell and the crystal. Heat treatment of kaolin at 800 or 950 °C resulted in the loss of the crystal structure and the formation of mullite and cristobalite and restoration of haemolytic activity. Mányai et al. (1970) established that the haemolytic activity of the kaolin samples is determined by the crystal structure and hydration of the surface. Studies by Adamis & Krass (1991) showed that the haemolytic activities of kaolin samples coming from different sources and used in the ceramic industry were significantly different. Narang et al. (1977) found that the haemolytic activity of the kaolin sample increased with the amount of silicic acid that could be dissolved from the particles.

Oscarson et al. (1981) determined the haemolytic activity of montmorillonite, kaolinite, quartz, and several fibrous silicates. Montmorillonite had the highest haemolytic activity, followed in decreasing order by quartz, palygorskite, chrysotile, and kaolinite. Polyvinyl-pyridine-*N*-oxide inhibited the haemolysis caused by montmorillonite and kaolinite. If the incubation medium contained sucrose instead of physiological saline, no haemolysis was observed with the dusts. These results indicate that the hydrogen bonding and ionic interactions are important in the haemolytic action of the silicate surface on the erythrocyte membrane.

Robertson et al. (1982) studied the effect of nitrous oxide on the cytotoxicity of dusts. Kaolinite adsorbed nitrous oxide; although the

Table 17. Toxicity of kaolin and illite clay *in vitro*

System / species / gender	Dose (mg/ml) / treatment[a]	Findings[a]	References
Peritoneal macrophages			
Rat (Sprague-Dawley CFY)	Two types of kaolin (90% kaolinite, 4% quartz; or 93% kaolinite, 4% quartz; diameter not given) incubated with cells for 24 h either as such or after dry milling for 32 h	Dry milling decreased the methylene blue adsorption to a third and the inhibition of TTC reduction by one-half to two-thirds.	Juhasz et al. (1978)
Mouse (Swiss T.O.)	Kaolin (non-specified), 100 µg/ml, incubation for 18 h	Approximately 25% release of LDH and β-glucuronidase release with native kaolin; one-half to two-thirds of the activity lost upon calcination.	Brown et al. (1980)
Rat (Sprague-Dawley CFY), male	Six different kaolins, kaolinite content 51–95%, quartz content 5–20%; 1.0 mg/ml (<5 µm diameter, median 1–2 µm) incubated in cell suspension for 1 h	All samples considered cytotoxic based on TTC reduction, except one (30%), which had 71% kaolinite and 22% quartz. All considered inert based on small LDH release, except the one with the highest quartz concentration (29%); the kaolinite concentration in this specimen was 67%.	Adamis & Timár (1976); Timár et al. (1979, 1980); Adamis et al., (1986); Adamis & Krass (1991)
	One sample of an illite clay (28% illite, 28% quartz); 1.0 mg/ml (<5 µm diameter, median 1–2 µm) incubated in cell suspension for 1 h	Cytotoxic based on TTC reduction, but not cytotoxic based on small LDH release.	

Table 17 (Contd)

System / species / gender	Dose (mg/ml) / treatment[a]	Findings[a]	References
Rat (Sprague-Dawley CFY), male	Four different kaolins (from Hungary, silica content 4, 5, 18, and 30%, <5 µm diameter, otherwise not specified) incubated in cell suspension for 24 h; similar experiment with a non-specified illite clay	Three out of four kaolins and illite considered cytotoxic based on TTC reduction; no relationship between quartz concentration and cytotoxicity. The clays studied did not induce release of LDH, but illite decreased intracellular LDH activity.	Adamis & Timár (1980); Adamis et al. (1985)
Mouse (T.O.), female	Culture of unstimulated macrophages with Cornwall kaolinite (not specified) for 18 h	Kaolinite induced LDH release from macrophages; this was prevented by polyvinylpyridine-N-oxide.	Davies & Preece (1983)
Mouse (T.O.), female	Cornwall kaolin (98% kaolinite, 2% mica), 98% <5 µm in diameter, incubated with cells at 40 µg/ml for 18 h	Kaolinite induced a 70% LDH release to the medium; the release was partly prevented by treatment of kaolin with polyvinylpyridine-N-oxide and fully prevented by additional treatment with polyacrylic acid.	Davies (1983)
Rat (Wistar, SPF), both sexes	Kaolin (composition not specified, diameter 0.2–25 µm), 0.5 mg/10[6] cells incubated for 2 h	Of all cells with particles, 0.6% and 1.6% dead cells with particles at 1 and 2 h (lowest toxicity group of three).	Styles & Wilson (1973)
Mouse	Fifteen respirable dust samples from kaolin drying and calcining plants in England (kaolinite content 84–96%, mica 3–6%, quartz 1%, feldspar 0–7%), a sample of Cornish kaolin (K1, 98% kaolinite, no quartz or feldspar, 2% mica), and a sample of Georgia kaolin (K2, 99% kaolinite, no	All dust samples were cytotoxic. The quartz content could not explain the cytotoxicity. The kaolinite samples showed a dose-dependent cytotoxicity, which could not be explained by their content of ancillary materials. Polyacrylic acid treatment of kaolin has only a	Davies et al. (1984)

Table 17 (Contd)

System / species / gender	Dose (mg/ml) / treatment[a]	Findings[a]	References
	quartz, mica, or feldspar, and reference quartz DQ12, mica, gibbsite, and titanium dioxide as controls; incubation for 18 h with macrophages, LDH release measured	small effect on its cytotoxicity, indicating that the positive charge at the edge of the mineral (blocked by acrylic acid) is not a major determinant of the toxicity.	
Alveolar macrophages			
Rat (Wistar, SPF), both sexes	Kaolin (composition not specified, diameter 0.2–25 µm), 0.5 mg/10^6 cells incubated for 2 h	Of all cells with particles, 3.7% and 4.2% dead cells with particles at 1 and 2 h (lowest toxicity group of three).	Styles & Wilson (1973)
Rabbit (New Zealand)	Kaolinite (>99% pure), >99% respirable size, incubated with cells at 0.25–2.5 mg/ml	Kaolinite caused an inhibition of amino acid incorporation into protein in a dose-dependent manner, 65% inhibition at 1 mg/ml. Inhibition reversed by addition of serum.	Low et al. (1980)
Guinea-pig	Kaolin (non-specified), 100 µg/ml, incubation for 18 h	Approximately 30% release of LDH and β-glucuronidase release with native kaolin; >90% of the activity lost upon calcination.	Brown et al. (1980)
Rabbit (New Zealand)	Kaolin (unspecified) and illite clay (unspecified) (<5 µm diameter), 0.5 mg/ml incubated in cell suspension for 24 h	Kaolin induced a 15.3% release of LDH and a 7% release of alkaline phosphatase. Illite clay induced a 2% release of LDH and a 1.3% release of alkaline phosphatase. For quartz, the figures were 51% and 16%.	Daniel & Le Bouffant (1980)

Table 17 (Contd)

System / species / gender	Dose (mg/ml) / treatment[a]	Findings[a]	References
Rat (Sprague-Dawley), male	Georgia kaolin (≥96% kaolinite, no quartz, >95% >5 μm in diameter), incubated with cells for 1 h at 0.1–1 mg/litre	Kaolin induced a dose-dependent release of LDH, β-glucuronidase, and β-N-acetylglucosaminidase of 60–80%. The effect was largely (9–15% release) abolished by lecithin.	Wallace et al (1985)
Rat (strain not specified)	Kaolin, 1.0 mg/ml (<5 μm diameter, MMAD 2.1 μm) incubated in cell suspension for 2 h	Kaolin induced an 80% release of LDH and a 60% release of β-glucuronidase and β-N-acetylglucosaminidase, being most cytotoxic of all minerals studied, quartz included.	Vallyathan et al. (1988)
Rat (Wistar), male	Alveolar macrophages incubated with Mexicali dust (see Table 13); LDH release measured	Concentration- and time-dependent release of LDH, which reacted 50% at 0.5 mg/ml and was much more pronounced than with quartz.	Osornio-Vargas et al. (1991)
Leukocytes			
Human phagocytic cells from one donor	Well crystallized standard kaolinite KGa-1, no quartz, cristobalite, or mica, particle size 3.2 μm median volume diameter	Kaolinite at concentrations of approximately 1 mg/ml induced luminol-dependent chemiluminescence as an expression of generation of reactive oxygen species in both monocytes and neutrophils when opsonized and when not opsonized.	Gormley et al. (1985)

Table 17 (Contd)

System / species / gender	Dose (mg/ml) / treatment[a]	Findings[a]	References
Erythrocytes			
Human washed erythrocytes	Erythrocytes incubated with Hungarian water-cleaned kaolin (composition not indicated; <5 μm in diameter), as such or after heat treatment at 290–900 °C for 90 min	Kaolin was strongly haemolytic; heating for 90 min to 200 or 350 °C increased, but heating to 500 or 650 °C practically abolished, the haemolytic potency. Kaolin heated at 800 or 950 °C was at least as potent a haemolyser as non-treated kaolin.	Mányai et al. (1970)
Sheep, plasma-free erythrocytes	Kaolin (source and composition unspecified), <5 μm in diameter, incubated with cells at 1 mg/ml for 2 h	Kaolin caused 40% haemolysis; acid and alkali treatments of the clay decreased its haemolytic potency.	Narang et al. (1977)
Rat (strain not specified)	Kaolin (unspecified) and illite clay (unspecified) (<5 μm in diameter) 1.0 mg/ml incubated in cell suspension for 1 h	Kaolin caused 98% haemolysis, illite 24% (quartz 48%).	Daniel & Le Bouffant (1980); Le Bouffant et al. (1980)
Rabbit	Kaolin (non-specified), incubation for 50 min	Twenty per cent haemolysis caused by 1.3 mg kaolin in a total volume of 4 ml; 11.6 mg of calcined kaolin was needed for the same effect.	Brown et al. (1980)
Sheep (citrated blood)	Kaolin (median volume diameter 4.7 μm) incubated in cell suspension	Haemolysis 60% in 30 min; haemolytic potency similar to that of quartz.	Ottery & Gormley (1978)
Sheep	Kaolin, 1.0 mg/ml (<5 μm diameter, MMAD 2.1 μm) at five concentrations, 0.1–1 mg/ml, incubated in cell suspension for 50 min	Linear, dose-dependent haemolysis; about 20% haemolysis with 0.5 mg/ml; 35% with 1 mg/ml; approximately 2 times as potent as quartz.	Vallyathan et al. (1988)

Table 17 (Contd)

System / species / gender	Dose (mg/ml) / treatment[a]	Findings[a]	References
Sheep (citrated blood, single donor)	Kaolinite 90% <2 μm in diameter at six concentrations, 0.005–0.25 mg/ml, incubated in cell suspension for 1 h	More haemolytic per mg than silica or talc; 20% haemolysis with 0.25 mg/ml; 95% with 25 mg/ml.	Woodworth et al. (1982)
Sheep	Georgia kaolin (≥96% kaolinite, no quartz, >95% >5 μm in diameter), incubated with cells for 1 h	Kaolin induced a dose-dependent haemolysis, which reached 42% at 1 mg/ml. The effect was completely abolished by lecithin.	Wallace et al. (1985)
Bovine erythrocytes washed with buffered saline	South Carolina kaolinite (composition not specified) incubated in cell suspension for 1 h	A concentration of 0.6 mg/ml induced 50% haemolysis of erythrocytes. Polyvinylpyridine-N-oxide, a hydrogen-bonding material, partly inhibited the haemolysis.	Oscarson et al. (1981)
Bovine erythrocytes washed with buffered saline	South Carolina kaolinite (composition not specified) at several concentrations of different particle sizes incubated in cell suspension for 1 h	A concentration of 0.6 mg/ml induced 50% haemolysis of erythrocytes. Of the minerals studied, kaolinite was least haemolytic, the potency being 1/20 that of silica. Particles of 0.2–2 μm diameter were most active; particles <0.2 μm or >20 diameter had no or little haemolytic activity; reduction of surface charge and cation exchange capacity by coating particles with an aluminium-hydroxy polymer largely eliminated haemolytic capacity.	Oscarson et al. (1986)

Table 17 (Contd)

System / species / gender	Dose (mg/ml) / treatment[a]	Findings[a]	References
Human (citrated blood)	Six different kaolins, kaolinite content 51–95%, quartz content 5–20%; 1.0 mg/ml (<5 µm diameter, median 1–2 µm) incubated in cell suspension for 1 h	Haemolysis 60–90% for all samples except one (30%), which had 71% kaolinite and 22% quartz.	Adamis & Krass (1991)
	One sample of an illite clay (28% illite, 28% quartz); 1.0 mg/ml (<5 µm diameter, median 1–2 µm) incubated in cell suspension for 1 h	Haemolysis 95%.	
Human	One "bentonite," containing 50% illite, 25% montmorillonite, 25% quartz; two undefined illite clays, one kaolinite with dickite and nakrite as main components and quartz as a minor component, one kaolin with kaolinite as the main component, and two unspecified kaolins ground in a ball mill to diameter <5 µm, incubated in cell suspension for 1 h	Fifty per cent haemolysis caused by 1.5–4 mg/ml kaolinites (0.06–0.115 m^2/ml); and 1.0–4.0 mg/ml (0.039–0.12 m^2/ml) illites. Haemolytic activity roughly proportionate to surface area of mineral powder; haemolytic activity largely lost after heating to over 500 °C.	Mányai et al. (1969)
Human red blood cells from normal donors	Dust from the town Mexicali (see Table 14)	Concentration of 2 mg/ml produced a 95 ± 3% haemolysis in 1 h; the haemolysis was stronger than with quartz.	Osornio-Vargas et al. (1991)

Table 17 (Contd)

System / species / gender	Dose (mg/ml) / treatment[a]	Findings[a]	References
Neural cells			
Neuroblastoma (N1E-115) cells with differentiation induced by dimethylsulfoxide	Standard kaolinite KGa-1 0.1–1.0 mg/ml incubated in cell suspension	Within minutes, resting potential depolarized and ability to maintain action potentials in response to stimulation was lost; within 30 min, severe morphological deterioration of cells.	Banin & Meiri (1990)
Neuroblastoma (N1E-115) cells and oligodendroglial (ROC-1) cells	South Carolina kaolinite (non-specified, mainly 1–2 µm diameter) incubated at 0.1 mg/ml in cell suspension for 24 h	No alteration of LDH activity in medium for either cell type; no decrease of the viability (assessed by trypan blue exclusion) of N1E-115 cells after 24 h.	Murphy et al. (1993b)
Other cell types and *in vitro* **systems**			
Tracheal epithelial (cloned cell line from Syrian hamster, strain 87.20) in log growth phase in monolayer	Kaolinite 90% <2 µm in diameter at four concentrations, 0.003–0.1 mg/ml, incubated in cell suspension for 24 h	Cells phagocytized clay particles; dose-dependent damage to plasma membrane as evidenced by loss of ^{51}Cr from cells; loss of ^{51}Cr after 24 h approximately 40% with 0.1 mg/ml, twice that of quartz.	Woodworth et al. (1982)
Human umbilical vein endothelial cells	South Carolina kaolinite (non-specified, mainly 1–2 µm diameter) incubated at 0.1 mg/ml in cell suspension for 24 h	Kaolinite induced a statistically significant 50% increase in LDH activity in the medium and killed 90% of the cells in 24 h.	Murphy et al. (1993b)

Table 17 (Contd)

System / species / gender	Dose (mg/ml) / treatment[a]	Findings[a]	References
Macrophage-like cell line P338D₁	Three kaolinites, 1 with "high crystallinity," 1 with "medium crystallinity," and 1 with "low crystallinity," diameter <5 μm, incubated at 80 μg/ml for 48 h	Kaolinites caused a 78–91% decrease in the viability of the cells and induced leakage of LDH and β-N-acetylglucosaminidase. Adsorption of nitrous oxide on the minerals slightly decreased the effect on viability.	Robertson et al. (1982)
V79-4 Chinese hamster lung cell line	Non-specified kaolin incubated with the cells for 6–7 days	LD_{50} 20 mg/ml for kaolin, which was the most toxic of the 21 particulate and fibrous materials tested.	Brown et al. (1980)
Macrophage-like cell line P338D₁	Thirty respirable dust specimens from coal mines in United Kingdom; cytotoxicity index developed from effects on trypan blue exclusion, release of LDH, glucos-aminidase, and lactic acid production	A positive correlation between ash content and cytotoxicity of the dusts. In dusts with >90% coal, there was also a correlation between kaolin + mica content and cytotoxicity.	Gormley et al. (1979)
Macrophage-like cell line P338D₁	Two kaolinites (KGa-1, KGa-2) from Source Clays repository, particle sizes 3.2 and 3.9 μm, with no cristobalite or quartz, incubated for 48 h	Cell viability not changed at 20 μg/ml, and 60–70% at 80 μg/ml.	Gormley & Addison (1983)
Isolated human leukocyte elastase	Cornwall kaolinite and four different illite clays (composition and particle size not specified), 5 μg/ml or 20 μg/ml	Kaolinite (5 μg/ml) caused 90% inhibition of the enzyme, illites (20 μg/ml), 10–53% inhibition.	Oberson et al. (1996)

Table 17 (Contd)

System / species / gender	Dose (mg/ml) / treatment[a]	Findings[a]	References
Artificial organelles			
Liposomes (artificial phospholipid membrane vesicles[b] 0.1–2 μm diameter) entrapping dissolved chromate (CrO_4^{2-})	Kaolinite, 90% <2 μm in diameter, at five concentrations, 0.1–10 mg/ml, incubated in cell suspension for 1 h	Dose-dependent loss of chromate from vesicles; loss of chromate after 1 h (in excess of spontaneous rate) approximately 20% with 10 mg/ml; spontaneous rate was 4–6%.	Woodworth et al. (1982)

[a] LD_{50} = median lethal dose; LDH = lactate dehydrogenase; MMAD = mass median aerodynamic diameter; TTC = 2,3,5-triphenyltetrazolium chloride.
[b] Prepared from dipalmitoyl phosphatidylcholine, sphingomyelin, cholesterol, and dicetylphosphate.

adsorption was small, it caused a significant decrease in the cytotoxicity of the dust.

6.6.2.2   Macrophage studies

Styles & Wilson (1973) studied the effect of a large number of mineral dusts and polymers on the alveolar and peritoneal macrophages of rat. The kaolin sample did not damage the macrophages. Five out of six kaolin samples decreased TTC reduction by peritoneal macrophages, but only one of them also induced a release of LDH to the medium (Adamis & Timar, 1976, 1986; Timar et al., 1979, 1980; Adamis et al., 1986; Adamis & Krass, 1991) (Table 17). The surface cationic groups may play a role in the cytotoxic effect of aluminium silicates (kaolin, illite, bentonite). This is suggested by the observed decrease in the cytotoxic effect of aluminium silicate by adsorption of paraquat cations (Adamis & Tímár, 1976). The cytotoxic effect increases with increasing adsorption capacity of the kaolin sample (Adamis & Timár, 1980; Adamis et al., 1985). This relationship has also been observed in connection with methylene blue, paraquat cation, serum protein, serum lipid, and serum phospholipid adsorption. After dry grinding the kaolin sample in an oscillating mill for 32 h, the adsorption and, at the same time, the macrophage cytotoxic effect decreased (Juhasz et al., 1978). The above indicates the importance of the nature of the crystal surface and the magnitude of the adsorption capacity on the sample's cytotoxicity.

Quartz, asbestos, and kaolinite increased the elastase secretion of peritoneal macrophages. The increase, however, was not as great as that observed after phagocytosis of latex particles. With alveolar macrophages, a difference was also found between the inert latex and the dust samples. This difference probably reflects the cytotoxicity of the dust. Kaolinite has a cytotoxic effect on both peritoneal and alveolar macrophages (White & Kuhn, 1980).

Davies (1983) and Davies et al. (1984) studied the cytotoxicity of Cornwall (United Kingdom) and Georgian (USA) kaolinite on mouse peritoneal macrophages. The Cornwall sample contained 98% kaolinite and 2% mica, and 98% of the particles were smaller than 5 μm in diameter; the Georgia kaolin contained 99% kaolinite and no quartz, mica, or feldspar. Both caused cytoplasmic LDH liberation from macrophages, and the cytotoxicity was apparently caused by kaolinite, not the other dust components. After polyvinylpyridine-*N*-oxide adsorption, the cytotoxic effect of kaolinite specimens from Cornwall decreased significantly (Davies & Preece, 1983; Davies et al., 1984). Davies et al. (1984) concluded that the amorphous silica-rich gel coating the kaolinite particles was probably responsible for the cytotoxic effect of kaolinite. The flake-like morphology of kaolinite is unlikely to play a role.

Low et al. (1980), studying rabbit alveolar macrophage cells, found that the presence of kaolinite inhibited the incorporation of amino acids into proteins through the non-competitive inhibition of amino acid transport.

Brown et al. (1980) compared different macrophage experimental systems. Kaolin proved to be cytotoxic; after thermal treatment (calcined kaolin), however, the damaging effect decreased significantly.

6.6.2.3 *Other tissue cultures and* in vitro *systems*

Gormley et al. (1979) studied the cytotoxicity of 30 respirable mine dust specimens with varying compositions from the United Kingdom using a macrophage-like P388 D1 cell line. Positive correlation was found between cytotoxicity and the kaolin and mica (combined) content of dusts containing >90% coal, but not in dusts containing less coal. In the study described in Table 16 on bentonite, Gormley & Addison (1983) studied the cytotoxicity of two standard kaolinites, KGa-1 and KGa-2, with particle sizes of 3.2 and 3.9 μm, which do not contain quartz or cristobalite. They showed very little effect on cell viability at 20 μg/ml; even at 80 μg/ml, the cell viability remained at 60% or more.

Kaolinite treated with hydrochloric acid did not increase the liberation of LDH from macrophages. The activity of quartz samples was considerably increased by treatment with hydrochloric acid (Kriegseis et al., 1987).

Bruch & Rosmanith (1985) found that mixed dusts containing kaolinite, illite, and quartz inhibit TTC reduction and induce liberation of LDH from guinea-pig alveolar macrophages; no apparent relationship emerged between the toxicity and dust composition.

6.6.3 *Other clays*

Illite inhibited TTC reduction but caused only a small LDH release from peritoneal macrophages and was haemolytic to human erythrocytes *in vitro* (Table 17).

6.6.4 *Relationships between* in vivo *and* in vitro *studies*

Tímár et al. (1980) did not find any direct relationship between the *in vitro* cytotoxicity of mineral dusts and their *in vivo* fibrogenicity. Mixed dusts containing 10–30% quartz that were inert in *in vitro* experiments were found to cause fibrosis *in vivo* after a long exposure time. Le Bouffant et al. (1980) reached the same conclusion based on *in vitro* (haemolysis, macrophage) and *in vivo* animal experiments (lung weight, collagen content after 3 and 12 months). On the other hand, of several particulate organic and inorganic materials, asbestos and polyurethane foam dust were the most toxic to alveolar and peritoneal macrophages and induced the most pronounced fibrosis in rat peritoneum; most of the materials studied neither were cytotoxic *in vitro* nor caused peritoneal fibrosis (Styles & Wilson, 1973).

6.7 Summary of the effects of quartz[1]

Quartz dust induces cellular inflammation *in vivo*. Short-term experimental studies of rats have found that intratracheal instillation of quartz particles leads to the formation of discrete silicotic nodules in rats, mice, and hamsters. Inhalation of aerosolized quartz particles impairs alveolar macrophage clearance functions and leads to progressive lesions and pneumonitis. Oxidative stress (i.e., increased formation of hydroxyl radicals, reactive oxygen species, or reactive nitrogen species) has been observed in rats after intratracheal instillation or inhalation of quartz. Many experimental *in vitro* studies have found that the surface characteristics of the crystalline silica particle influence its fibrogenic activity and a number of features related to its cytotoxicity. Although many potential contributory mechanisms have been described in the literature, the mechanisms responsible for cellular damage by quartz particles are complex and not completely understood.

---

[1] This section has been taken directly from IPCS (2000), pages 4–5.

Long-term inhalation studies of rats and mice have shown that quartz particles produce cellular proliferation, nodule formation, suppressed immune function, and alveolar proteinosis. Experimental studies of rats reported the occurrence of adenocarcinomas and squamous cell carcinomas after the inhalation or intratracheal instillation of quartz. Pulmonary tumours were not observed in experiments with hamsters or mice. Adequate dose–response data (e.g., no-adverse-effect or lowest-adverse-effect levels) for rats or other rodents are not available because few multiple-dose carcinogenicity studies have been performed.

Quartz did not test positively in standard bacterial mutagenesis assays. Results of genotoxicity studies of quartz conflict, and a direct genotoxic effect for quartz has not been confirmed or ruled out.

In experimental studies of particles, results may vary depending on the test material, particle size of the material, concentration administered, and species tested. The experiments with quartz particles involved specimens from various sources, using various doses, particle sizes, and species, which could have affected the observations.

Data on the reproductive and developmental effects of quartz in laboratory animals are not available.

7. EFFECTS ON HUMANS

7.1 General population

General population exposure to low concentrations of clay minerals is universal (see section 4.1). However, no studies are available on the health effects of clays in the general population.

7.2 Occupational exposure

There are important limitations in the data on the health effects of clays in humans. Most studies are case reports or case series (bentonite) or cross-sectional in design (kaolin). Smoking data are not available in most of the studies. In many studies, exposure characterization, both qualitatively and quantitatively, is incomplete.

7.2.1 Bentonite

Data on the effects of occupational exposure to bentonite are summarized in Table 18.

Five workers exposed to bentonite or Fuller's earth (Campbell & Gloyne, 1942; Tonning, 1949; Sakula, 1961; Gibbs & Pooley, 1994) for periods ranging from less than 15 years to more than 40 years developed dyspnoea, in some cases accompanied by cough and sputum, and pneumoconiosis. The general characteristics of the pneumoconiosis of the above five workers, as revealed on autopsy, included emphysema, lesions, numerous firm black patches and nodules 1–20 mm in diameter, dilated air sacs, adhesions, isolated cavities filled with "black sludge," and other evidence of physical deterioration. Gibbs & Pooley (1994) reported that the accumulated dust amounted to 128 mg/g of dry lung (analysed by transmission electron microscopy). There was very little or no fibrosis of the type associated with silicosis. Microscopic examination of lung tissue revealed that the nodules were aggregates of fine brown pigment lying free in tissue spaces or contained in macrophages held in a fine reticulum. The pigment was confirmed as Fuller's earth by microscopic examination and, in the studies of Sakula (1961) and Gibbs & Pooley (1994), by X-ray diffraction. Authors were unanimous that the pathology was markedly

Table 18. Case reports and case series on effects of occupational exposure to bentonite[a]

Mineral	Exposure	Effects	References
Fuller's earth[b], montmorillonite 85–90%, silica trace	1 worker, 38 years to dust from processing (milling)	Increasing dyspnoea and cough prior to abrupt death after illness at age 56; no tuberculosis; autopsy: pulmonary fibrosis, soft adhesions, and black fibrotic patches, but no silicotic nodules; bronchiole had thickened walls with collagenous fibres containing dust particles.	Campbell & Gloyne (1942)
Fuller's earth,[c] montmorillonite[d]	2 workers, 10 and 14 years to high but unmeasured concentrations of dust from grinding, drying, and bagging	Radiological evidence of damage to lungs, including nodules and thickened areas; respiratory problems.	Gattner (1955)
	3 workers, 10 months to 3 years to dust	Radiological evidence for possible slight damage to lungs.	
	3 workers, 0.5–15 years to dust	No radiological evidence for damage to lungs.	
Fuller's earth,[e] calcium mont-morillonite, 0.8% quartz	1 worker, >40 years to dust from processing (drying and bagging); also smoked cigarettes	Shortness of breath; bilateral, fine reticulonodular shadowing in X-ray; autopsy (after haemorrhage from gastric ulcer): lungs — pneumoconiosis; emphysema; soft, grey black, stellate lesions 4–5 mm across; dust (montmorillonite)-laden macrophages around bronchioles; 128 mg mineral dust/g dry tissue analysed by transmission electron microscopy.	Gibbs & Pooley (1994)

Table 18 (Contd)

Mineral	Exposure	Effects	References
Fuller's earth,[d] montmorillonite with some quartz and amorphous silica	49 workers; average dust concentrations 18, 32, 56, 120, and 390 mg/m[3] in different parts of the plant 5 years prior to medical study, and apparently higher prior to these measurements	Twenty-eight workers with exposures of 3.5–16 years had significant radiological evidence of damage to lungs. Duration of employment was 5.6, 7.7, 10, and 16 years for the workers whose pulmonary damage was classified as 1, 2, 3, or 4 on a scale of 1–4. The number of workers in these categories was 21, 18, 8, and 2.	McNally & Trostler (1941)
Fuller's earth[b]	5 workers, 4–39 years to dust working in grinding and sieving of Fuller's earth	Exposure 4, 5, or 19 years, no radiological evidence of damage to lungs; 35 and 39 years, radiological evidence of damage to lungs; at autopsy, one of the latter two showed advanced pneumoconiosis with dark nodules and extensive patchy fibrosis, but no evidence of silicotic nodulation or massive fibrosis.	Middleton (1940)
Bentonite	180 workers in a bentonite mine and 73 and 39 workers in two bentonite refining factories; bentonite contained 30% free silicon dioxide as vitreous silica and β-cristobalite; particle size was 54%, 62%, and 95% ≤1 μm in diameter	Prevalence of silicosis in the examined groups was 6% in the mine, 35.5% in factory A with high exposure, and 12.8 in factory B. Pulmonary effects were linked more to intensity than to duration of exposure in different sections of the refinery.	Rombola & Guardascione (1955)

Table 18 (Contd)

Mineral	Exposure	Effects	References
Bentonite[f] with 5–11% silica	32 workers, for varying periods to dust from processing; dust concentration at four plants 2–275 mppcf (average of 17 samples, 76 mppcf), no other known exposure to silica dust	Fourteen workers showed radiological evidence of silicosis, ranging from minimal to advanced.	Phibbs et al. (1971)
	1 worker for 10 years to dust; no other known exposure to silica dust	Death at age 72, apparently from advanced silicosis.	
	2 workers, ages 62 and 67, for 10 years to dust; no other known exposure to silica dust	Radiological evidence of silicosis; 67-year-old had lung nodules 1–3 cm in diameter, with high silica content (4.6 mg/1.5 g tissue) and chronic granulomatous inflammation.	
	All the workers (exact number not specified) in a bentonite processing mill, for varying periods to varying concentrations of dust	Eight (all regular cigarette smokers) showed radiological evidence of pulmonary fibrosis ranging from scattered nodules (1 person) and fine nodules (1 person) to progressive massive fibrosis (3 persons).	
Fuller's earth[b]	1 worker for 42 years to dust at a high but unmeasured concentration	Pneumoconiosis; chest radiographs showed fine to medium mottling of lungs; no evidence of tuberculosis; died at age 67 (no autopsy).	Sakula (1961)

Table 18 (Contd)

Mineral	Exposure	Effects	References
Fuller's earth[b]	1 worker for 35 years to dust at a high but unmeasured concentration	Pneumoconiosis; chronic cough and sputum with increasing dyspnoea; chest radiographs showed fine mottling of lungs; no evidence of tuberculosis; autopsy after death at age 65 showed lungs emphysematous with many 1- to 2-mm round, black nodules containing montmorillonite and possibly other mineral matter derived from montmorillonite.	Sakula (1961)
Fuller's earth[b]	1 worker for an unknown period (thought to be less than 15 years) to dust	Death from pneumonia at age 79 (about 50 years after last exposure); autopsy revealed pneumoconiosis, emphysema, and numerous small, black, firm nodules 5–7.5 mm in diameter in the lungs; nodules were aggregates of a fine brown pigment consisting mainly of macrophages containing dust particles composed mainly of Fuller's earth; no evidence of fibrosis.	Tonning (1949)

[a] All cases of pulmonary exposure described below apparently involved only male workers.
[b] Nutfield district, Surrey, United Kingdom.
[c] Sachen-Anhalt, Germany; in part of the exposure period for workers with long exposures, 5–20% quartz sand was added to the Fuller's earth during processing.
[d] Illinois, USA.
[e] Bath, United Kingdom.
[f] Wyoming, USA.
[g] mppcf = million particles per cubic foot; 30 mppcf equals approximately 10 mg/m^3.

different from that present in silicosis. McNally & Trostler (1941) summarized the observations of Middleton (1940), who reported pathology similar to the above in a worker exposed to Fuller's earth dust for 35 or 39 years.

The observations of Tonning (1949) support the conclusions of Ungváry et al. (1983) that clearance of bentonite stored in foci from the lung is relatively slow. Bentonite stored in nodules in the lungs that were noted on autopsy stemmed from exposure to dust more than 50 years previous to death at age 79.

Three of the studies summarized in Table 18 (Middleton, 1940; Gattner, 1955; Sakula, 1961) reported radiological evidence, not supported by autopsy, of significant damage to lungs following exposures to bentonite or montmorillonite dust of 10–42 years. Gattner (1955) also reported radiological evidence of possible slight damage to lungs in three workers following 1–3 years of exposure and, in three other workers, no radiological evidence of damage following 0.5–15 years of exposure. Dust concentrations were described as high by Sakula (1961), but were not measured in any of these studies. McNally & Trostler (1941) reported significant radiological evidence of lung damage in 28 workers exposed to bentonite dust for 3.5–16 years, but severity of damage was not correlated with period of exposure or the age of the individual. Twenty-one co-workers with exposures of 1.75–16 years, however, displayed no radiological evidence of damage. The authors ascribed this lack of correlation between damage and period of exposure to the fact that dust concentrations varied widely within the plant (Table 18) and that the men worked at different jobs during the course of their employment and thus were exposed to varying concentrations of dust, which overwhelmed any effects from total period of employment. No attempt was made in any of the above studies to approximately reconstruct actual exposures, as has been done, for example, with silica exposure (Rice et al., 1984). Mikhailova-Docheva et al. (1986), however, reported that damage to the lungs increased with increasing exposure to bentonite dust (containing 1% quartz), although radiological data on Bulgarian workers presented in their paper appeared too limited to permit any final conclusions to be drawn. Their data concurred with those of McNally & Trostler (1941) on the absence of radiological evidence of lung damage in many exposed workers. The data also supported the reasonable assumption of greater damage from higher concentrations of dust. Bulgarian processing workers involved in grinding and packaging bentonite and exposed to 20–40 mg/m^3 dust displayed in general more severe lung damage than production workers involved in removing bentonite from open pit mines and exposed to 1–5 mg/m³ dust.

Observations by Phibbs et al. (1971) demonstrate that exposure to bentonite dust containing 5–11% silica can lead to silicosis (Table 18). Exposure to this bentonite dust for 10 years with no other known exposure to silica dust led to advanced silicosis in three workers and to death, apparently from silicosis, in one of these three. All showed radiological evidence of silicosis, and the presence of silica in the lungs was confirmed by biopsy in one case. In a sample of 32 workers expo
sed for varying periods to dust from processing, 14 showed radiological evidence of silicosis (Phibbs et al., 1971). These observations suggest that the beneficial effects ascribed to bentonite by Timár et al. (1966) and by Havrankova & Skoda (1993) in protecting rat lungs from silicosis may not be present in humans.

The role of silica in inducing pneumoconiosis in all of the above studies on bentonite and Fuller's earth workers remains uncertain. All of the mineral deposits contained at least trace amounts of silica, and, in the case of the workers examined by Gattner (1955), some were exposed during a period in which quartz sand was added to the Fuller's earth in processing.

The effects of the confounding factor tobacco smoking on the progressive damage to workers' lungs from inhaling bentonite dust remain unknown. Two of the papers cited in Table 18, Phibbs et al. (1971) and Gibbs & Pooley (1994), mention that some of the afflicted workers smoked cigarettes, but other papers provide no information on this matter. In view of the ability of cigarette smoke to aggravate the development of pneumoconiosis in asbestos miners (Blanc & Gamsu, 1989), dental technicians (Choudat, 1994), and coal miners (Sadler & Roy, 1990), level of smoking may be an important factor in determining the degree of lung damage from bentonite exposure.

7.2.2 Kaolin

Many case reports and case series have suggested that exposure to kaolin causes pneumoconiosis (Lemaistre, 1894; Hlava, 1897; Middleton, 1936; Perry, 1947; Thomas, 1952; Lynch & McIver, 1954; Hale et al., 1956; Edenfield, 1960; Warraki & Herant, 1963; Sheers, 1964; Vallyathan et al., 1982; Levin et al., 1996). In several cases, however, it was not clear whether kaolinite and quartz or quartz alone was responsible for the resulting pneumoconiosis (Gudjonsson & Jacobsen, 1934; Sundius et al., 1936).

7.2.2.1   Kaolin workers, United Kingdom

In England, a number of papers have dealt with the effects of kaolin originating from Cornwall mines (Table 19). Hale et al. (1956) performed medical and radiological examinations of those occupationally exposed to kaolin and reported in detail on six workers who had been working in the drying and bagging of kaolin. All had radiological pneumoconiosis, and two were further studied in autopsy. In one case, characteristic silicotic-type nodulation together with progressive tuberculosis were found. Large quantities of kaolinite and amorphous quartz were found in the lung. In another case, large quantities of pure kaolinite (as much as 20–40 g) were found in the lung without tuberculosis but with severe fibrosis. The disease was like the pneumoconiosis of coal miners and differed from classic silicosis. In the upper part of the lung, greyish or blue-greyish massive confluent lesions were described, which were not as hard on palpation as the silicotic conglomerates. Hale and co-workers (1956) also studied the lung of a kaolin worker from Georgia, USA, and observed that the dust in the lungs consisted entirely of kaolinite; no trace of quartz could be seen.

In a cross-sectional study of 533 Cornish china clay workers, there was an exposure time-dependent increase in the prevalence of radiologically diagnosed kaolinosis, from 4% in those with less than 15 years of exposure to 19% among those with more than 25 years of exposure (Sheers, 1964). Among 526 workers with less than 5 years of exposure, no kaolinosis was observed. Among the workers who were more heavily exposed (milling and bagging), the prevalence was 6% for those with a work history of 5–15 years and 23% for those with an exposure of more than 15 years. Confluent lesions were found in 12 workers, and 30 workers had International Labour Office (ILO) categories 2–3 lesions (ILO, 1959). There was little evidence of disability related to kaolinosis; only one worker with massive fibrosis had become disabled and changed to a lighter job.

Table 19. Effects of occupational exposure to kaolin and other clays on health

Study design, studied population	Exposure measurement	Exposure	Findings[a]	References
Cross-sectional; 533 Cornish china clay workers	Occupational history from records for everyone. Millers, baggers, loaders considered to be continuously exposed, kiln workers and drymen intermittently exposed	Exposure to china clay; industrial hygiene expertise used to group occupations based on the intensity of exposure; no quantitative data or qualitative assessment of the dust	Exposure time-dependent increase in the prevalence of radiologically diagnosed kaolinosis, from 4% in those with less than 15 years of exposure to 19% among those with more than 25 years of exposure. Among 526 workers with less than 5 years of exposure, no kaolinosis was observed. Among the workers who were more heavily exposed (milling and bagging), the prevalence was 6% for those with a work history of 5–15 years and 23% for those with an exposure of >15 years. Confluent lesions were found in 12 workers, and 30 workers had ILO categories 2–3 lesions. There was little evidence of disability related to kaolinosis; only one worker with massive fibrosis had become disabled and changed to a lighter job.	Sheers (1964)
Cross-sectional; 1728 Cornish china clay workers in 1977	Occupational group and history of work in each group for all workers	Exposure to china clay; industrial hygiene expertise used to group occupations based on the intensity of exposure; no quantitative data or qualitative assessment of the dust	77.4% of workers in pneumoconiosis category 0, 17.9% in category 1, 4.7% in category 2 or 3. Advanced pneumoconiosis in 19 workers. Every dusty job contributed to the amount of simple pneumoconiosis. Smoking unrelated to radiographic appearance. Vital capacity deteriorated with advancing pneumoconiosis; for "FEV," the association was not statistically significant. Subjective symptoms not related to past exposure as assessed by years worked in different jobs.	Oldham (1983)

Table 19 (Contd)

Study design, studied population	Exposure measurement	Exposure	Findings[a]	References
Cross-sectional; 3831 employees and 336 pensioners in china clay industry in United Kingdom in 1985	Analysis by job classification	Average exposures at the time of the study 0.5–2.7 mg/m³ (see Table 9)	3374 workers in pneumoconiosis category 0, 271 in category 1, 39 in category 2, and 5 in category 3. Employment in mills had strongest effect on pneumoconiosis category, followed by dryers (in keeping with the exposure levels). In kilns, the exposure was also high, but pneumoconiosis was not equally prevalent; kaolin there is no longer crystalline. Work in mills or as dryer before 1971 had twice the effect on pneumoconiosis prevalence as work there after 1971. Ventilatory capacity related to radiological status, no consistent independent relation with occupational history was observed. Respiratory symptoms were related to ventilatory function.	Ogle et al. (1989)
Cross-sectional; 4401 current and retired china clay workers in United Kingdom in 1990	Dust measurements from 1978, estimated exposure before 1978; detailed occupational history for each participant	Average respirable dust exposure 1.2–4.7 mg/m³ (see Table 9)	Small opacity profusion related to work dustiness and total occupational dust dose; to reach category 1 by age 60, the estimated total dose for non-smokers was 85 mg/m³·years, for smokers, 65 mg/m³·years. The major determinant of respiratory symptoms was smoking; total dust exposure had a minor effect.	Rundle et al. (1993)

Table 19 (Contd)

Study design, studied population	Exposure measurement	Exposure	Findings[a]	References
Cross-sectional; 4401 current and retired china clay workers in United Kingdom in 1990	Dust measurements from 1978, estimated exposure before 1978; detailed occupational history for each participant	Average respirable dust exposure 1.2–4.7 mg/m^3 (see Table 9)	Univariate analysis showed relationships between lung function and age, X-ray score, smoking class, occupational history, and total occupational dust dose. In multiple regression analysis, when the effects of age, X-ray score, and smoking class had been accounted for, there was no independent additional effect from total occupational dust dose or occupational history.	Comyns et al. (1994)
Cross-sectional; 39 current and 16 ex-kaolin workers at a kaolin mine and mill in Georgia, USA	Respirable and total dust analysed at the time of study (see Table 9)	Dust composed of 96% kaolinite, 4% titanium dioxide, no free silica, no asbestiform fibres; average respirable dust in all job categories ≤2 mg/m^3	Of current and ex-workers with ≥5 years of exposure ($n = 55$), 4 had simple pneumoconiosis and 4 complicated pneumoconiosis. Mean adjusted FVC, FEV$_1$, peak flow lower ($P < 0.05$) among kaolin workers than among 189 non-kaolin-exposed referents.	Sepulveda et al. (1983)
Cross-sectional; 459 workers in three Georgia kaolin mining and processing facilities with ≥1 year work history; mean	Details on measurements not given	At the time of study, US Mine Safety and Health Administration documented exposure to kaolin dust <5 mg/m^3 with less than 1% free silica; survey	417 workers had pneumoconiosis category 0, 29 category 1, 8 category 2, and 5 category 3. Of the blacks, 13.6%, and of the whites, 7.6% had pneumoconiosis. Pneumoconiosis was significantly related to age, >15 years of exposure, and greatest dust exposure. Complicated pneumoconiosis (large opacities) related to decreased respiratory function,	Kennedy et al. (1983)

Table 19 (Contd)

Study design, studied population	Exposure measurement	Exposure	Findings[a]	References
duration of employment 12 years		in one of the plants in 1951 and 1960 showed kaolin dust concentrations of 377 and 361 mg/m³, in 1951, the raw material had 0.25% free silica	but otherwise there was no correlation between pneumoconiosis and respiratory function.	
Cross-sectional; all 65 men employed in a Georgia kaolin mine studied	During 5-year period, 157 measurements of respirable dust (see Table 9)	Dust composed of 94–98% kaolinite and 2–6% anastase (TiO₂); no asbestiform fibres of crystalline silica; mean respirable dust in processing area 1.74 mg/m³, 0.14 mg/m³ in the mine	Five of the workers had radiological pneumoconiosis. All had worked in the processing area. For the whole group, FVC and FEV₁ were within the normal range, but they were lower for the workers with pneumoconiosis. FVC and FEV₁ decreased with years of employment in the processing area. Pneumoconiosis was not related to smoking.	Altekruse et al. (1984)
Cross-sectional; 2379 current kaolin workers in Georgia, USA	No measurements; occupational title as proxy	The authors state that the free silica exposure is negligible because of washing out of impurities in the process	4.4% prevalence of category ≥1 simple pneumoconiosis, 0.89% prevalence of complicated pneumoconiosis; 7.1% of white and 19% of black dry processors and 4.1% of white and 9.1% of white wet processors had pneumoconiosis. FEV₁ was <80% of the expected among 7.5, 12.8, and 33.3% and FVC was <80% among 8.0, 10.5, and 33.3% of those with normal chest radiograph and those with simple and complicated pneumoconiosis. Similarly, among	Morgan et al. (1988)

Table 19 (Contd)

Study design, studied population	Exposure measurement	Exposure	Findings[a]	References
			lifelong non-smokers, the frequency of lowered FEV_1 and FVC was elevated among those with complicated pneumoconiosis.	
Cross-sectional: workers of two Georgia, USA, kaolin plants	Reanalysis of data from two of the three plants studied by Kennedy et al. (1983); no measurements		19/162 and 21/223 workers in the two plants had pneumoconiosis. The adjusted prevalence of pneumoconiosis increased 1.1% for each year in production. Workers in plant 1 had a 2.7 times higher prevalence than workers in plant 2. In plant 1, 15–20% of production had been calcined kaolin, while plant 2 had produced hydrous clay only.	Baser et al. (1989)
Cross-sectional; 914 workers in earthenware industry, United Arab Republic	No quantitative measurements	83–86% of respirable dust potassium aluminium silicate, 1–2% free silica	264 workers has worked for <10 years, 133 for 10–15 years; no pneumoconiosis cases were observed in these groups. 326 workers had worked 15–20 years and included 4 cases with pneumoconiosis, and 191 workers had worked >20 years and included 2 with pneumoconiosis. 2 of the pneumoconiosis cases were classified as progressive massive fibrosis; 4 had dyspnoea on exertion.	Warraki & Herant (1963)

Table 19 (Contd)

Study design, studied population	Exposure measurement	Exposure	Findings[a]	References
Cross-sectional, 11 workers in the bagging section of a kaolin refinery in Sri Lanka, average age 35.2 years, duration of employment 3–9 years (average 6 years)	No measurements available	Kaolinite content of kaolin >99%, 76.8% <3 μm in diameter	No radiological abnormalities observed. The absence of pneumoconiosis most likely due to short duration of employment and young age of the workers studied.	Uragoda & Fernando (1974)
Cross-sectional; 18 factories in heavy clay industry in United Kingdom; 1934 current employees	1465 personal dust samples collected at the time of the study and analysed for respirable dust and quartz	Cumulative exposure <40 mg/m³·years respirable dust for 94% and <4 mg/m³·years quartz for 93%; for details, see Table 10	1.4% had radiographic small opacities in ILO category ≥1/0, 0.4% had small opacities in ILO category ≥2/1. Risk dependent on lifetime exposure to quartz and respirable dust. Anamnestic chronic bronchitis, wheezing, and breathlessness related to dust exposure. Dust and quartz exposure strongly correlated.	Love et al. (1999)

Table 19 (Contd)

Study design, studied population	Exposure measurement	Exposure	Findings[a]	References
Cross-sectional; 268 current brick workers in South Africa	97 respirable, 78 total dust, and 29 silica analyses from three of the five participating factories at the time of the study	Mean respirable dust and total dust exposure 2.22 and 15.6 mg/m^3; mean free silica 2.1%; for exposure in different groups, see Table 10	4.2% had profusion score ≥1/0; profusion score significantly related to cumulative respirable dust exposure. Anamnestic respiratory symptoms and FVC and FEV$_1$ significantly related to exposure to respirable dust.	Myers & Cornell (1989); Myers et al. (1989b)

[a] FEV = forced expiratory volume; FEV$_1$ = forced expiratory volume in 1 s; FVC = forced vital capacity; ILO = International Labour Organization.

In a further questionnaire, a radiological and lung function study among 1676 workers of the largest china clay producers in Cornwall, advanced pneumoconiosis was observed in 19 workers, all employed in dusty workplaces, and pneumoconiosis category 2 or 3 was reported in 79 workers. Vital capacity showed a significant reduction with increasing pneumoconiosis. No symptoms related to years worked in different jobs were observed (Oldham, 1983).

Wagner et al. (1986) examined the lungs from 62 workers who had died between 1968 and 1981, had been employed in kaolin production, and had been referred to the pneumoconiosis unit of the Medical Research Council of the United Kingdom as suspected of having kaolin-induced lung damage. In lung specimens from 16 of the workers, $\geq 90\%$ of the lung dust was kaolinite, $\leq 1.1\%$ quartz, and $\leq 1.0\%$ feldspar; none of these workers had worked with china stone. Among the 16 workers in whose lung specimens there was $<90\%$ kaolinite, $\geq 0.9\%$ quartz, and $\geq 1.1\%$ feldspar, 12 had worked with china stone. Nodular fibrosis appeared to be related to a high quartz content of the dust recovered from the lung, whereas among those with a high content of kaolinite in the lungs, interstitial fibrosis was observed.

In a cross-sectional study of employees and retirees from the United Kingdom china clay industry, 271 of 3374 workers had ILO pneumoconiosis category 1, 39 category 2, and 5 category 3 (ILO, 1981). Employment in dusty operations was accompanied by the highest risk of pneumoconiosis; however, in kilns where exposure was also high, pneumoconiosis was less frequent. As the authors propose, the likely explanation is that in the kiln department, kaolinite is no longer crystalline. Ventilatory capacity was related to the radiological status of the lungs (Ogle et al., 1989).

In a further follow-up of the United Kingdom china clay workers, small opacity profusion was related to work dustiness and estimated total occupational dust dose. The major determinant of respiratory symptoms was smoking; total dust exposure had only a minor effect (Rundle et al., 1993). In a further analysis of the same workforce, univariate analysis demonstrated a relationship between lung function and age, X-ray score, smoking class, occupational history, and total occupational dust dose. In multiple regression analysis, total occupational dust dose and occupational history were not independent additional risk factors for decreased lung function (Comyns et al., 1994).

7.2.2.2   Kaolin workers, USA

Kaolin workers in Georgia have been studied extensively (Table 19). Edenfield (1960) found pneumoconiosis in 44 (3.9%) of 1130 persons working with kaolin. All these had worked for a number of years in the loading area or some other area of the plant that in earlier years had been very dusty. Only 2 had worked in the kaolin industry less than 10 years, and 19 had been employed for more than 20 years. All of those classed as having severe pneumoconiosis (stage III) had worked more than 20 years, and all had worked in heavy dust in the car loading and bagging areas. Thirty-one of the 44 had a stage I pneumoconiosis; they had no symptoms or signs of respiratory dysfunction. Similarly, the seven workers with stage II pneumoconiosis exhibited no respiratory symptoms. The six cases with stage III pneumoconiosis also had emphysema, had complaints of cough and dyspnoea, and had been placed in jobs requiring minimal activity.

Lapenas & Gale (1983) found diffuse reticulonodular lung infiltration and a nodule in the upper lung lobe in a 35-year-old worker of a Georgia kaolin processing factory who had been occupationally exposed to kaolin aerosol for 17 years. Exploratory thoracotomy revealed an $8 \times 12 \times 10$ cm conglomerate pneumoconiotic lesion containing large amounts of kaolinite. Quartz could not be demonstrated by scanning electron microscopy or X-ray diffraction.

Lapenas et al. (1984) performed a pathological study on biopsy or autopsy specimens from five patients admitted to the Medical Center Hospital of central Georgia between 1976 and 1981, who were estimated to represent some of the most advanced cases of kaolin pneumoconiosis seen in that hospital. Respiratory failure was a contributing factor to death in two of the three autopsy cases. Chest X-ray demonstrated small irregular shadows and large obscure patches typical of kaolin pneumoconiosis. Histological examinations revealed significant kaolinite deposits and peribronchial nodules. The nodules differed from those in silicotic patients and consisted mainly of kaolinite aggregates traversed by fibrous tissue trabecules. The presence of kaolinite in the lungs was confirmed by mineralogical examinations, while quartz could not be demonstrated.

Sepulveda et al. (1983) examined 39 current and 16 ex-workers of a Georgia kaolin mine and mill. Average respirable dust levels were approximately 0.2 mg/m^3 in the mine and between 1 and 2 mg/m^3 for other work stations. Samples of respirable dust contained 96% kaolinite, 4% titanium dioxide, and no silica. Pneumoconiosis was found in 15% of the workers and ex-workers with 5 years or more of exposure. Kaolin workers had a decreased forced vital capacity (FVC), forced expiratory volume in 1 s (FEV$_1$), and mean adjusted peak flow when the values were adjusted for age, height, race, and pack-years of smoking ($P < 0.05$).

Kennedy et al. (1983) examined 459 kaolinite workers from central Georgia with a mean duration of employment of 12 years. Pneumoconiosis occurred in 9.2% of the workers and, among employees older than 54 years, was associated with high dust exposure. Among black workers with "complicated pneumoconiosis," the FVC was 81.6%, indicating borderline restriction, and the FEV$_1$/FVC was significantly lowered. The raw material from the mine contained 0.25% free crystalline silica. At the time of the study, the airborne concentrations of kaolin dust were below 5 mg/m^3 and the dust contained less than 1% free silica, but dust concentrations up to 377 mg/m^3 had been recorded in the past.

In a cross-sectional study of a Georgia kaolin mine and processing plant, Altekruse et al. (1984) examined the pulmonary function and lung radiography of all 65 employees. The respirable dust concentration at the time of the study was 0.14 mg/m^3 in the mine area and 1.74 mg/m^3 in the kaolinite processing area, but had been higher earlier (see Table 9). About 94–98% of the particles were kaolinite and 2–6% anastase (titanium dioxide). Asbestiform fibres and crystalline silica were not present in the samples. Five workers, all of whom worked in the kaolinite processing area with highest exposure, showed radiological evidence of pneumoconiosis; they had worked for the company for 7–36 years. There was a slight, exposure duration-dependent decrease in the FVC and FEV$_1$ among the workers ($P < 0.05$), those with radiological pneumoconiosis having lower lung function than the others.

In a cross-sectional study of pulmonary function and radiology among workers of 12 kaolin companies in Georgia (Morgan et al., 1988), an increased prevalence of pneumoconiosis and decreased lung function (FEV$_1$, but not FVC) were observed among the workers working with calcined clay. Among workers with more than a 3-year tenure, the adjusted prevalence of simple and complicated pneumo-coniosis was 3.2% and 0.63%, respectively. Smoking did not explain the presence of decreased FEV, nor was there a relationship between pneumoconiosis and lung function. The authors stated that the workers were not exposed to quartz.

7.2.2.3   Kaolin workers, other countries

Warraki & Herant (1963) examined radiologically 914 china clay workers of Ayyat, United Arab Republic; 264 of them had been exposed for less than 10 years, 133 for 10–15 years, 326 for 15–20 years, and 191 for more than 20 years. Pneumoconiosis was diagnosed in six workers, all with exposures longer than 15 years. In two cases, confluent masses were found in the lungs; during a 2.5-year follow-up, one of these died with cor pulmonale. No measurements of dust were reported; a sample of airborne dust contained 1–2% free silica (Table 19).

Uragoda & Fernando (1974) examined workers in kaolinite processing in Sri Lanka. No individuals working with wet clay were examined because their risk of disease from dust exposure was considered low. Among the 11 persons working with dry materials (sacking and weighing), X-rays showed no sign of disease. The most frequent complaints were skin irritation and dermatitis, probably due to the tropical climate (high temperature and humidity). Lack of pneumoconiosis was to be expected because of the short exposure time (average 6 years, range 3–9 years).

7.2.2.4   Brick and tile workers and others

Lesser et al. (1978) described nine cases of disabling pneumoconiosis in workers who had worked in the Missouri, USA, fire brick industry. The chest X-ray and lung tissue analysis (from two cases) were consistent with silicosis.

In an extensive cross-sectional study of the respiratory health of brick workers in South Africa (Myers & Cornell, 1989; Myers et al., 1989b), 4.2% of the workforce studied had a radiological profusion score $\geq 1/0$, and it was significantly related to cumulative respirable dust exposure. Similarly, anamnestic respiratory symptoms and respiratory function measurements (FVC and FEV_1) were significantly related to exposure to respirable dust (Table 19).

Love et al. (1999) performed a cross-sectional study on the respiratory health of workers in the heavy clay industry in the United Kingdom and analysed the results with relation to dust exposure at the time of the study. Of the workforce studied, 1.4% had radiographic small opacities in the ILO category $\geq 1/0$, and 0.4% had radiographic small opacities in the ILO category $\geq 2/1$ (ILO, 1981). The risk of radiological abnormality and of anamnestic chronic bronchitis, wheezing, and breathlessness increased with estimated lifetime exposure to respirable dust and quartz. The exposures to dust and quartz correlated strongly with each other (Table 19).

Seaton et al. (1981) described findings for four pneumoconiosis cases referred to the Pneumoconiosis Medical Board of the United Kingdom for disablement benefit and who for practically all their working life had worked in shale mining. Pneumoconiosis was identified in two cases at autopsy, in one by lobectomy, and in one by X-ray. Two of the patients were found at autopsy also to have peripheral squamous lung cancer. The clinical and histological picture of the disease mainly resembled the pneumoconiosis of coal miners and kaolin workers. The lung dust was analysed for three of the cases and was found to contain kaolinite (19–36%), mica (13–16%), and quartz (9–18%).

7.3   Summary of the effects of quartz[1]

There are many epidemiological studies of occupational cohorts exposed to respirable quartz dust. Silicosis, lung cancer, and pulmonary tuberculosis are associated with occupational exposure to quartz dust. IARC [the International Agency for Research on Cancer] classified inhaled crystalline silica (quartz or cristobalite) from occupational sources as a Group 1 carcinogen based on sufficient evidence of carcinogenicity in humans and experimental animals; "in making the overall evaluation, the Working Group noted that carcinogenicity in humans was not detected in all industrial circumstances studied. Carcinogenicity may be dependent on inherent characteristics of the crystalline silica or on external factors affecting its biological activity or distribution of its polymorphs" (IARC, 1997b).

---

[1] This section is taken directly from IPCS (2000), page 5.

Statistically significant increases in deaths or cases of bronchitis, emphysema, chronic obstructive pulmonary disease, autoimmune-related diseases (i.e., scleroderma, rheumatoid arthritis, systemic lupus erythematosus), and renal diseases have been reported.

Silicosis is the critical effect for hazard identification and exposure–response assessment. There are sufficient epidemiological data to allow the risk of silicosis to be quantitatively estimated, but not to permit accurate estimations of risks for other health effects mentioned above.

The risk estimates for silicosis prevalence for a working lifetime of exposure to respirable quartz dust concentrations of about 0.05 or 0.10 mg/m^3 in the occupational environment vary widely (i.e., 2–90%).[1] Regarding exposure to ambient quartz in the general environment, a benchmark dose analysis predicted that the silicosis risk for a continuous 70-year lifetime exposure to 0.008 mg/m^3 (estimated high crystalline silica concentration in US metropolitan areas) is less than 3% for healthy individuals not compromised by other respiratory diseases or conditions and for ambient environment (US EPA, 1996b). The silicosis risk for persons with respiratory diseases exposed to ambient quartz in the general environment was not evaluated.

---

[1] For the development of silicosis, it has been estimated that 45-year exposure to 0.05 mg/m^3 quartz leads to a prevalence of 2–30% of ILO category ≥1/1 silicosis; for an exposure of 0.10 mg/m^3, the predicted range of the prevalence of ILO category ≥1/1 silicosis is 3–90%.

8. EFFECTS ON OTHER ORGANISMS IN THE LABORATORY AND FIELD

Information on environmental effects associated with bentonite, kaolin, or other clays is limited.

In laboratory experiments utilizing both genetically engineered microorganisms and their normal counterparts, increasing concentrations of montmorillonite in the soil enhanced the rate of nitrification (Jones et al., 1991).

The 96-h LC_{50} for rainbow trout (*Oncorhynchus mykiss*) of Wyoming bentonite, used as a viscosifier in drilling fluids, was 19 g/litre (Sprague & Logan, 1979).

A commercial viscosifier described as finely ground sodium bentonite[1] did not kill hardhead catfish (*Galeichthys [Arius] felis*), striped mullet (*Mugil cephalus*), pigfish (*Orthopristis chrysoptera*), sheepshead (*Archosargus probatocephalus*), silver perch (*Bairdiella chrysoura*), red drum (*Sciaenops ocellatus*), Atlantic croaker (*Micropogonias undulatus*), pinfish (*Lagodon rhomboides*), southern kingfish (*Menticirrhus americanus*), black drum (*Pogonias cromis*), crevalle jack (*Caranx hippos*), spot (*Leiostomus xanthurus*), Atlantic spadefish (*Chaetodipterus faber*), planehead filefish (*Stephanolepis hispida*), Atlantic midshipman (*Nautopaedium [Porichthys] porosissimus*), gulf killifish (*Fundulus grandis*), sheepshead minnow (*Cyprinodon variegatus*), striped burrfish (*Chilomycterus schoepfi*), puffer (*Sphoeroides* spp., unspecified), searobins (*Prionotus* spp., unspecified), coneshell (*Penaceus* spp., unspecified), blue crab (*Callinectes sapidus*), or shell oyster (*Ostrea virginica*) at the highest concentration tested, 7.5 g/litre, within 24 h. One spotted seatrout (*Cynoscion nebulosus*) (number tested not given) died at 6 g/litre in 22 h (Daugherty, 1951).

The 24- and 48-h LC_{50} values for kaolinite toxicity to the water flea (*Daphnia pulex*) were >1.1 g/litre (Lee, 1976).

---

[1] http://www.baroididp.com/baroid_idp_prod/baroid_idp_prod_datasheets.asp

Georgia kaolin caused <10% mortality of sea urchin (*Strongylocentrosus purpuratus*), Japanese clam (*Tapes japonica*), hermit crab (*Pagurus hirsutiusculus*), isopod (*Sphaeroma pentodon*), mud snail (*Nassarius obsoletus*), blue mussel (*Mytilus edulis*), and tunicates (*Molgula manhattensis* and *Styela montereyensis*) within 5–12 days. The 200-h LC_{10} values for coast mussel (*Mytilus californianus*), black-spotted bay shrimp (*Crangon nigromaculata*), migrant prawn (*Palaemon macrodactylus*), dungeness crab (*Cancer magister*), and the polychaete *Neanthes succinea* were 26, 16, 24, 10, and 9 g/litre, respectively. The 100-h LC_{10} values for the tunicate *Ascidia ceratodes*, amphipod *Anisogammarus confervicolus*, and shiner perch (*Cymatogaster aggregata*) were 7, 38, and 1 g/litre, respectively (McFarland & Peddicord, 1980).

No effect on the hatching success or egg development rate of four marine fish species — red seabream (*Pagrus major*), black porgy (*Acanthopagrus schlegeli*), striped knifefish (*Oplegnathus fasciatus*), and threeline grunt (*Parapristipoma trilineatum*) — was observed at kaolinite concentrations up to 10 g/litre for 24 h. Larvae were more sensitive to kaolinite: the 12-h LC_{50} values were 170 and 710 mg/litre for *P. trilineatum* and *O. fasciatus*, respectively; mortality was also observed for *P. major* at concentrations of 1000 mg/litre and above (Isono et al., 1998).

In a study on the effects of drilling fluid components on soils and plants, Miller et al. (1980) dosed natural soil (pH 6, with 8.1% organic matter and 34.7% moisture content) with bentonite. No effect was observed on the growth of beans (*Phaseolus vulgaris*) or corn (*Zea mays*) at a concentration of 135 g/1.6 kg soil.

In a series of 100 necropsies among 11 mammalian and eight avian species in the San Diego Zoo, California, USA, interstitial fibrosis was observed in 20% of the mammals studied. In birds, granulomas around tertiary bronchi, but no fibrosis, were observed. Electron and X-ray diffraction showed that the minerals present in the lungs were muscovite mica and illite (Brambilla et al., 1979). As the authors noted, the finding does not infer causality between clays present in the lungs and pulmonary effects.

9. EVALUATION OF HUMAN HEALTH RISKS AND EFFECTS ON THE ENVIRONMENT

9.1 Evaluation of human health risks

Bentonite, kaolin, and other clay materials vary remarkably in composition, depending on the source and process of manufacture. This is a major obstacle for any universal hazard characterization for all these products.

Several elements and other chemicals may be adsorbed to clays because of their cation exchange capacity and, if released, may exert toxic effects.

One of the important components of all clays is crystalline silica, the amount of which varies greatly. Crystalline silica causes various lung diseases, including silicosis and lung cancer. The content of crystalline silica will often be the decisive factor in clay-induced adverse health effects. For quartz-induced lung cancer, no reliable quantitative risk estimates can be made. For quantitative estimates of quartz-induced silicosis, see below and chapter 7.

There are no studies on the possible adverse effects of clay minerals upon direct skin contact; however, both bentonite and kaolin are used extensively in cosmetics, and the absence of reports of adverse effects indicates that these clays pose no important health hazards via the dermal route.

There are no studies in humans of the effects of long-term oral exposure to clays (or clays contaminated with toxic chemicals). No adequate studies on the carcinogenicity of bentonite were available. One limited study on kaolin in experimental animals did not demonstrate important adverse long-term toxicity or carcinogenicity.

No data are available on the genotoxicity of clay materials *in vitro* or *in vivo* (except for quartz). No information is available on the developmental toxicity of clay materials or their effects on fertility in humans; one very limited study on the developmental toxicity of bentonite and another on kaolin did not indicate adverse effects on the progeny.

Information on the toxicity of bentonite by the inhalation route in humans is limited to case reports and case series, in most of which both qualitative information and quantitative information on exposure are very limited and confounding factors, such as exposure to silica and tobacco smoke, have not been considered.

From the limited data available from studies on bentonite-exposed persons, retained montmorillonite appears to effect only mild non-specific tissue changes, which are similar to those that have been described in the spectrum of changes of the "small airways mineral dust disease" (nodular peribronchiolar dust accumulations containing refractile material [montmorillonite] in association with limited interstitial fibrosis). In some of the studies, radiological abnormalities have also been reported.

There exist no reported cases of marked diffuse/nodular pulmonary tissue fibrotic reaction to montmorillonite in the absence of free silica.

Bentonite is toxic to a variety of mammalian cells *in vitro*. When introduced to tissues, it produces transient, local inflammation. Bentonite increases susceptibility to bacterial infection in mice. No such effect has been reported in humans exposed to bentonite dust.

Many reports of well studied cases and a substantial number of cross-sectional studies, mainly in the United Kingdom and Georgia, USA, have clearly demonstrated that long-term exposure to kaolin may lead to a pneumoconiosis, which has been named kaolinosis. This disease develops even in the absence of exposure to silica, as demonstrated by the absence of silica in the dust and in lung tissue specimens. The disease is usually mild. Symptoms and lowered pulmonary function are observed only when the radiological changes are prominent.

A quantitative universal estimate of the risk of kaolinosis cannot be made with the data available. However, based on data from United Kingdom china clay workers (exposed to clay containing <1.1% quartz), it has been estimated that to reach the ILO profusion category $\geq 1/1$ by age 60, the estimated cumulative exposure to kaolin for non-smokers is 85 mg/m^3·years, and for smokers, 65 mg/m^3·years (Rundle et al., 1993). For comparison, IPCS (2000) estimated that 45-year exposure to quartz at 0.05 mg/m^3 (i.e., a cumulative exposure of

2.25 mg/m^3·years) leads to a prevalence of 2–30% of ILO category ≥1/1 silicosis; for an exposure of 0.10 mg/m^3 (4.5 mg/m^3·years), the predicted range of the prevalence of ILO category ≥1/1 silicosis is 3–90%.

It can thus be very roughly estimated that kaolin is at least an order of magnitude less potent than quartz.

Kaolin is toxic to a variety of mammalian cells *in vitro*, and it produces transient inflammation in the lungs of experimental animals after intratracheal instillation.

Illite shows some cytotoxicity in *in vitro* studies and limited pulmonary toxicity in experimental animals after intratracheal instillation.

The decreasing rank order of the potencies of quartz, kaolinite, and montmorillonite to produce lung damage is consistent with their known relative "active" surface areas and surface chemistry.

9.2 Evaluation of effects on the environment

Bentonite and kaolin have low toxicity to aquatic species, a large number of which have been tested.

There is no reason to believe that the mining or processing of bentonite, kaolin, and other clays poses significant toxicological dangers to the environment. However, physical disturbance to the land, excessive stream sedimentation, and similar destructive processes resulting from the large-scale mining and processing of clays, like any large-scale mining operation, have a potential for significant environmental damage.

10. CONCLUSIONS AND RECOMMENDATIONS FOR PROTECTION OF HUMAN HEALTH AND THE ENVIRONMENT

Kaolin produces a specific pneumoconiosis, known as kaolinosis. Its fibrogenic potential is considered to be at least an order of magnitude less than that of quartz. Specific exposure limits should be set, and kaolin should not be considered an inert (nuisance) dust.

With regard to bentonite, a comparable montmorillonite pneumoconiosis has not been consistently reported. Based on its surface chemistry, lack of fibrogenicity in experimental systems, and limited human findings, inhaled bentonite is likely to be less dangerous to humans than kaolin.

In order to decrease adverse health effects from occupational exposure to clays, the Task Group recommends that the responsible risk managers:

- set limits for occupational exposure to clay materials, taking into consideration the quartz content;
- enforce and ascertain compliance with the limits by regular exposure monitoring;
- prepare guidelines to ensure good workplace practice;
- disseminate information on the hazards to exposed workers in an appropriate form;
- institute appropriate medical monitoring programmes to protect populations at risk (including pre-employment and periodical medical examinations, including, where appropriate, chest X-ray); and
- oblige clay producers to declare the content of hazardous components, such as quartz, of the saleable products.

For the protection of the general public, the Task Group recommends that when clays are used for medical and cosmetic purposes, not only the total content but also the mobility and bioavailability of potentially toxic substances in the products be established.

11. FURTHER RESEARCH

In order to improve the assessment of risks from exposure to clays, the Task Group recommends research in the following areas:

- contribution of clay mineral particles in the possible health effects of the fine and ultrafine particles in ambient air;
- longitudinal studies among occupationally exposed people with mineralogical analysis of the particles and the particle distribution in the air and, where possible, in the lung tissues of the deceased;
- studies on long-term effects and carcinogenicity in animals of pure minerals with defined particle size distribution by inhalation;
- experimental studies on inhaled clay minerals and interactions and possible synergism with tobacco smoke;
- genotoxicity studies *in vitro* and *in vivo*;
- studies on the possible uptake and kinetics of clay components after exposure via the gastrointestinal tract and the lungs;
- kinetics of adsorption to and release from clays of toxicants and nutrients;
- mechanistic studies for the identification of methods to predict the fibrogenicity of particulate matter from short-term studies; and
- potential use of different imaging methods in the determination of preclinical (early) stages of clay-induced pulmonary disease.

12. PREVIOUS EVALUATIONS BY INTERNATIONAL BODIES

Bentonite was evaluated by the Joint FAO/WHO Expert Committee on Food Additives in 1976, but no acceptable daily intake was allocated (JECFA, 1976).

REFERENCES

Adamis Z & Krass BK (1991) Studies on the cytotoxicity of ceramic respirable dusts using *in vitro* and *in vivo* test systems. Ann Occup Hyg, **35**: 469–483.

Adamis Z & Timár M (1976) Effects of various mineral dusts on macrophages *in vitro*. Int Arch Occup Environ Health, **37**: 301–307.

Adamis Z & Timár M (1978) Studies on the effect of quartz, bentonite and coal dust mixtures on macrophages *in vitro*. Br J Exp Pathol, **59**: 411–415.

Adamis Z & Timár M (1980) Investigations of the effects of quartz, aluminium silicates and colliery dusts on peritoneal macrophages. In: Brown RC, Gormley IP, Chamberlain M, & Davies R ed. The *in vitro* effects of mineral dusts. London, Academic Press, pp 13–18.

Adamis Z, Tátrai E, Timár M, & Ungváry G (1985) Evaluation of dust toxicity by short-term methods. In: Beck EG & Bignon J ed. *In vitro* effects of mineral dusts: 3rd international workshop. NATO Advanced Research Workshop on *In Vitro* Effects of Mineral Dusts, Schluchsee. Berlin, Springer-Verlag, pp 453–458 (NATO Advanced Science Institutes Series G3).

Adamis Z, Timár M, Köfler L, Tátrai E, & Ungváry G (1986) Biological effects of the respirable dusts from ore mines. Environ Res, **41**: 319–326.

Adamis Z, Tátrai E, Honma K, & Ungváry G (1998) Studies on the effect of quartz, bentonite and kaolin dust mixtures on rat lung. In: Chiyotami C, Hosoda Y, & Aizawa Y ed. Advances in the prevention of occupational respiratory diseases: proceedings of the 9th international conference on occupational respiratory diseases, Kyoto, 13–16 October 1997. Amsterdam, Elsevier Science B.V., pp 851–854.

Altekruse EB, Chaudhary BA, Pearson MG, & Morgan WK (1984) Kaolin dust concentrations and pneumoconiosis at a kaolin mine. Thorax, **39**: 436–441.

Ampian SG (1985) Clays. In: Mineral facts and problems. Washington, DC, US Bureau of Mines, pp 1–13 (Bulletin 675).

Attygalle D, Harrison CV, King EJ, & Mohanty GP (1954) Infective pneumoconiosis. I. The influence of dead tubercle bacilli (BCG) on the dust lesions produced by anthracite, coal-mine dust, and kaolin in the lungs of rats and guinea-pigs. Br J Ind Med, **11**: 245–259.

Bailey MR, Fry FA, & James AC (1982) The long-term clearance kinetics of insoluble particles from the human lung. Ann Occup Hyg, **26**: 273–290.

Bailey SW (1980a) Structures of layer silicates. In: Brindley GW & Brown G ed. Crystal structures of clay minerals and their X-ray identification. London, Mineralogical Society, pp 1–123 (Monograph No. 5).

Bailey SW, chairman (1980b) Summary of recommendations of AIPEA [Association Internationale Pour l'Étude des Argiles] nomenclature committee on clay. Am Mineral, **65**: 1–7. Available at: http://www.minsocam.org/msa/collectors_corner/arc/nomenclaturecl1.htm.

Banin E & Meiri H (1990) Toxic effects of alumino-silicates on nerve cells. Neuroscience, **39**: 171–178.

Bartley DL & Breuer GM (1982) Analysis and optimization of the performance of the 10 mm cyclone. Am Ind Hyg Assoc J, **43**: 520–528.

Baser ME, Kennedy TP, Dodson R, Rawlings W Jr, Rao NV, & Hoidal JR (1989) Differences in lung function and prevalence of pneumoconiosis between two kaolin plants. Br J Ind Med, **46**: 773–776.

Bates RE & Jackson JA ed. (1987) Glossary of geology, 3rd ed. Alexandria, Virginia, American Geological Institute, pp 65, 123, 262, 432.

Belt TH & King EJ (1945) Chronic pulmonary disease in South Wales coal miners. III. Experimental studies. D. Tissue reactions produced experimentally by selected dusts from South Wales coal mines. Med Res Counc Spec Rep Ser, 250: 29–68.

Bernstein M, Pairon J-C, Morabia A, Gaudichet A, Janson X, & Brochard P (1994) Non-fibrous dust load and smoking in dental technicians: a study using bronchoalveolar lavage. Occup Environ Med, **51**: 23–27.

Blanc PD & Gamsu G (1989) Cigarette smoking and pneumoconiosis: structuring the debate. Am J Ind Med, **16**: 1–4.

Boros DL & Warren KS (1970) Delayed hypersensitivity-type granuloma formation and dermal reaction induced and elicited by a soluble factor isolated from Schistosoma mansoni eggs. J Exp Med, **132**: 488–507.

Boros DL & Warren KS (1973) The bentonite granuloma. Immunology, **24**: 511–529.

Brambilla C, Abraham J, Brambilla E, Benirschke K, & Bloor C (1979) Comparative pathology of silicate pneumoconiosis. Am J Pathol, **96**: 149–170.

Briant JK & Moss OR (1984) The influence of electrostatic charge on the performance of 10-mm nylon cyclones. Am Ind Hyg Assoc J, **45**: 440–445.

Brindley GW & Brown G ed. (1980) Crystal structures of clay minerals and their X-ray identification. London, Mineralogical Society, 495 pp (Monograph No. 5).

Brody AR & Craighead JE (1975) Cytoplasmic inclusions in pulmonary macrophages of cigarette smokers. Lab Invest, **32**: 125–132.

Browett PJ, Simpson LO, & Blennerhassett JB (1980) Experimental granulomatous inflammation: the ultrastructure of the reaction of guinea pigs to bentonite injection. J Pathol, **130**: 57–64.

Brown RC, Chamberlain M, Davies R, Morgan DML, Pooley FD, & Richards RJ (1980) A comparison of 4 "in vitro" systems applied to 21 dusts. In: Brown RC, Gormley IP, Chamberlain M, & Davies R ed. The in vitro effects of mineral dusts. London, Academic Press, pp 47–52.

Bruch J & Rosmanith J (1985) Properties of mixed mine dusts and their relationship to pneumoconiosis. In: Beck AG & Bignon J ed. *In vitro* effects of mineral dusts: 3rd international workshop. NATO Advanced Research Workshop on *In Vitro* Effects of Mineral Dusts, Schluchsee. Berlin, Springer-Verlag, pp 433–440 (NATO Advanced Science Institutes Series G3).

Byers PD & King EJ (1959) Experimental infective pneumoconiosis with coal, kaolin and mycobacteria. Lab Invest, **8**: 647–664.

Campbell AH & Gloyne SR (1942) A case of pneumoconiosis due to the inhalation of Fuller's earth. J Pathol Bacteriol, **54**: 75–79.

Carleton HM (1924) The pulmonary lesions produced by the inhalation of dust in guinea-pigs. A report to the Medical Research Council. J Hyg, **22**: 438–478.

Carr DD, Rooney LF, & Freas RC (1994) Limestone and dolomite. In: Carr DD ed. Industrial minerals and rocks. Littleton, Colorado, Society for Mining, Metallurgy, and Exploration, pp 606–629.

Chariot P, Couste B, Guillon F, Gaucichet A, Bignon J, & Brochard P (1992) Non-fibrous mineral particles in bronchoalveolar lavage fluid and lung parenchyma from the general population. Am Rev Respir Dis, **146**: 61–85.

Cheng YS, Yamada Y, Yeh HC, & Swift DL (1988) Diffusional deposition of ultrafine aerosols in a human nasal cast. J Aerosol Sci, **19**: 741–751.

Choudat D (1994) Occupational lung diseases among dental technicians. Tubercle Lung Dis, **75**: 99–104.

Churg A (1983) Nonasbestos pulmonary mineral fibers in the general population. Environ Res, **31**: 189–200.

Churg A & Wiggs B (1985) Mineral particles, mineral fibers, and lung cancer. Environ Res, **37**: 364–372.

CIREP (2003) Final report [by the Cosmetic Ingredient Review Panel] on the safety of aluminium silicate, calcium silicate, magnesium aluminium silicate, magnesium silicate, magnesium trisilicate, sodium magnesium silicate, zirconium silicate, attapulgite, bentonite, Fuller's earth, hectorite, kaolin, lithium magnesium silicate, lithium magnesium sodium silicate, montmorillonite, pyrophyllite, and zeolite. Int J Toxicol, **22**(Suppl 1): 37–122.

Cohen IK, Diegelmann RF, & Wise WS (1976) Biomaterials and collagen synthesis. J Biomed Mater Res, **10**: 965–970.

Comyns RA, Ogle CJ, Rundle EM, Cockroft A, & Rossiter CE (1994) Dose–response for kaolin: The effect of kaolin on workers' health. Ann Occup Hyg, **38**(Suppl 1): 825–831.

Costabel U, Donner CF, Haslam PL, Rizzato G, Teschler H, Velluti G, & Wallaert B (1990) Occupational lung diseases due to inhalation of inorganic dust. In: Klech H & Hutter C ed. Clinical guidelines and indications for bronchoalveolar lavage (BAL): Report of the European Society of Pneumology Task Group on BAL. Eur Respir J, **3**(8): 946–949, 961–969.

Cronberg S & Caen JP (1971) Release reaction in washed platelet suspensions induced by kaolin and other particles. Scand J Haematol, **8**: 151–160.

Daniel H & Le Bouffant L (1980) Study of a quantitative scale for assessing the cytotoxicity of mineral dusts. In: Brown RC, Gormley IP, Chamberlain M, & Davies R ed. The *in vitro* effects of mineral dusts. London, Academic Press, pp 33–39.

Daugherty FM (1951) Effects of some chemicals used in oil well drilling on marine animals. Sewage Ind Wastes, **23**: 1282–1287.

Davies R (1983) Factors involved in the cytotoxicity of kaolinite towards macrophages *in vitro*. Environ Health Perspect, **51**: 249–252.

Davies R & Preece AW (1983) The electrophilic mobilities of minerals determined by laser Doppler velocimetry and their relationship with the biological effect of dusts toward macrophages. Clin Phys Physiol Meas, **4**: 129–140.

Davies R, Griffiths DM, Johnson NF, Preece AW, & Livingston DC (1984) The cytotoxicity of kaolin towards macrophages *in vitro*. Br J Exp Pathol, **65**: 453–466.

Deer WA, Howie RA, & Zussman J (1975) An introduction to rock-forming minerals. London, Longman Group Ltd., 528 pp.

Degueldre G (1983) Dust sampling. In: Parmeggiani L ed. Encyclopaedia of occupational health and safety, Vol 1. Geneva, International Labour Office, pp 689–695.

Dougherty SH, Fiegel VD, Nelson RD, Rodeheaver GT, & Simmons RL (1985) Effects of soil infection potentiating factors on neutrophils *in vitro*. Am J Surg, **150**: 306–311.

Dufresne A, Case B, Fraser R, & Perrault G (1994) Protocol of lung particulate analysis by electron transmission microscopy for decoding occupational history from lung retention. Ann Occup Hyg, **38**(Suppl 1): 503–517.

Dumortier P, de Vuyst P, & Yernault JC (1989) Non-fibrous inorganic particles in human bronchoalveolar lavage fluids. Scanning Microsc, **3**: 1207–1218.

Edenfield RW (1960) A clinical and roentgenological study of kaolin workers. Arch Environ Health, **1**: 392–403.

Eller PM & Cassinelli ME ed. (1994) Particulates not otherwise regulated, respirable 0600; and Particulates not otherwise regulated, total 0500. In: NIOSH manual of analytical methods, 4th ed. Cincinnati, Ohio, US Department of Health and Human Services, National Institute for Occupational Safety and Health.

Fan CM & Aw PC (1988) Processing of illite powder in Bidor, Perak: a study of the process and potential use of illite clay. Warta Geol, **15**(1): 37.

Fan CM & Aw PC (1989) Processing of illite powder in Bidor, Perak: a study of the process and the potential uses of illite clay. Bull Geol Soc Malaysia, **24**: 67–77.

Fanning DS, Keramides VZ, & El-Desoky MA (1989) Micas. In: Dixon JB & Weed SB ed. Minerals in soil environments. Madison, Wisconsin, Soil Science Society of America, pp 551–634.

Ferraro JR ed. (1982) The Sadtler infrared spectra handbook of minerals and clays. Philadelphia, Pennsylvania, Sadtler Research Laboratories, pp 354–356.

Fiore S, Cavalcante F, Medici L, Ragone PP, Lettino A, Barbaro M, Passariello B, & Quaresima S (2003) Trace element mobility in shales: Implications on geological and environmental studies. In: Proceedings of Euroclay 2003. 10th conference of the European Clay Groups Association, Modena, Italy, pp 99–100.

Gattner H (1955) Die Bleicherde-Lunge. Arch Gewerbepathol Gewerbehyg, **13**: 508–516.

Gaudette HE, Edes JL, & Grim RE (1966) The nature of illite. In: Bradley WF & Bailey SW ed. Clays and clay minerals. Oxford, Pergamon Press, pp 33–48.

Gerde P, Cheng Y-S, & Medinsky MA (1991) *In vivo* deposition of ultrafine aerosols in the nasal airway of the rat. Fundam Appl Toxicol, **16**: 330–336.

Gibbs AR & Pooley FD (1994) Fuller's earth (montmorillonite) pneumoconiosis. Occup Environ Med, **51**: 644–661.

Giese WW (1989) Countermeasures for reducing the transfer of radiocesium to animal derived foods. Sci Total Environ, **85**: 317–327.

Goldstein B & Rendall RE (1969) The relative toxicities of the main classes of minerals. In: Shapiro HA ed. Pneumoconiosis. Proceedings of the international conference, Johannesburg. Capetown, Oxford University Press, pp 429–434.

Gormley IP & Addison J (1983) The *in vitro* toxicity of some standard clay mineral dusts of respirable size. Clay Miner, **18**: 153–163.

Gormley IP, Collings P, Davis JM, & Ottery J (1979) An investigation into the cytotoxicity of respirable dusts from British collieries. Br J Exp Pathol, **60**: 523–536.

Gormley IP, Kowolik MJ, & Cullen RT (1985) The chemiluminescent response of human phagocytic cells to mineral dusts. Br J Exp Pathol, **66**: 409–416.

Grayson RL & Peng SS (1986) Characterization of respirable dust on a longwall panel. A case study. In: Khair AW ed. Engineering health and safety in coal mining. Proceedings of a symposium, Society for Mining, Metallurgy, and Exploration annual meeting, New Orleans, Louisiana. Littleton, Colorado, Society for Mining, Metallurgy, and Exploration, pp 95–117.

Greene-Kelly W (1955) Dehydration of montmorillonite minerals. Mineral Mag, **30**: 604–615.

Grim RE (1947) Differential thermal curves of prepared mixtures of clay minerals. Am. Mineral, **32**: 493–501.

Grim RE (1968) Clay mineralogy, 2nd ed. New York, McGraw-Hill, 596 pp.

Grim RE & Rowland RA (1944) Differential thermal analysis of clays and shales, a control and prospecting method. J Am Ceram Soc, **27**: 65–76.

Grim RE, Bray RH, & Bradley WF (1937) The mica in argillaceous sediments. Am Mineral, **22**: 813–829.

Gudjonsson SV & Jacobsen CJ (1934) A fatal case of silicosis. J Hyg, **34**: 166–171.

Hale LW, Gough J, King EJ, & Nagelschmidt G (1956) Pneumoconiosis of kaolin workers. Br J Ind Med, **13**: 251–259.

Hanchar JM, Stroes-Gascoyne S, & Browning L ed. (2004) Scientific basis for nuclear waste management. XXVIII. Materials Research Society symposium proceedings. Warrendale, Pennsylvania, Materials Research Society.

Hatch GE, Slade R, Boykin E, Hu PC, Miller FJ, & Gardner DE (1981) Correlation of effects of inhaled versus intratracheally injected metals on susceptibility to respiratory infection in mice. Am Rev Respir Dis, **124**: 167–173.

Hatch GE, Boykin E, Graham JA, Lewtas J, Pott F, Loud K, & Mumford JL (1985) Inhalable particles and pulmonary host defense: *in vivo* and *in vitro* effects of ambient air and combustion particles. Environ Res, **36**: 67–80.

Haukenes G & Aasen J (1972) Heterogeneity in the reactivity of antibodies with kaolin. Acta Pathol Microbiol Scand B, **80**: 251–256.

Haury BB, Rodeheaver GT, Pettry D, Edgerton MT, & Edlich RF (1977) Inhibition of non-specific defenses by soil infection potentiating factors. Surg Gynecol Obstet, **144**(1): 19–24.

Havrankova J & Skoda V (1993) Fibrogenic activity of artificial binary mixtures of mineral dusts and quartz studied on animals. In: Hurych J, Lesage M, & David A ed. Proceedings of the 8th international conference on occupational lung diseases. Prague, Czech Medical Society, pp 1231–1236.

Henderson RF, Damon EG, & Henderson TR (1978a) Early damage indicators in the lung. I. Lactate dehydrogenase activity in the airways. Toxicol Appl Pharmacol, **44**: 291–297.

Henderson RF, Muggenburg BA, Mauderly JL, & Tuttle WA (1978b) Early damage indicators in the lung. II. Time sequence of protein accumulation and lipid loss in the airways of beagle dogs with beta irradiation of the lung. Radiat Res, **76**: 145–158.

Hlava J (1897) Ein Fall von Silico-Aluminosis (Kaolinosis) der Lunge. Wiener Klin Rundschau, **11**: 611–612.

Hornfeldt CS & Westfall ML (1996) Suspected bentonite toxicosis in a cat from ingestion of clay cat litter. Vet Hum Toxicol, **38**(5): 365–366.

Hosterman JW & Patterson SH (1992) Bentonite and Fuller's earth resources of the United States. Washington, DC, US Geological Survey, 45 pp (USGS Professional Paper 1522).

Huertas FJ, Chou L, & Wollast R (1998) Mechanism of kaolinite dissolution at room temperature and pressure: Part 1. Surface speciation. Geochim Cosmochim Acta, **62**: 417–431.

IARC (1997a) Zeolites other than erionite. In: Silica, some silicates, coal dust and *para*-aramid fibrils. Lyon, International Agency for Research on Cancer, pp 307–333 (IARC Monographs on the Evaluation of Carcinogenic Risks to Humans, Vol 68).

IARC (1997b) Silica. In: Silica, some silicates, coal dust and *para*-aramid fibrils. Lyon, International Agency for Research on Cancer, pp 41–242 (IARC Monographs on the Evaluation of Carcinogenic Risks to Humans, Vol 68).

IARC (1997c) Palygorskite (attapulgite). In: Silica, some silicates, coal dust and *para*-aramid fibrils. Lyon, International Agency for Research on Cancer, pp 245–255 (IARC Monographs on the Evaluation of Carcinogenic Risks to Humans, Vol 68).

IARC (2001) Ionizing radiation, Part 2: Some internally deposited radionuclides. Lyon, International Agency for Research on Cancer, 595 pp (IARC Monographs on the Evaluation of Carcinogenic Risks to Humans, Vol 78).

ILO (1959) Meeting of experts on the international classification of radiographs of the pneumoconioses. Geneva, International Labour Office, pp 63–69.

ILO (1981) Guidelines for the use of ILO international classification of radiographs of pneumoconioses, 1980 ed. Geneva, International Labour Office.

Inouye S & Kono R (1972) Effect of a modified kaolin treatment on serum immunoglobulins. Appl Microbiol, **23**: 203–206.

IPCS (2000) Crystalline silica, quartz. Geneva, World Health Organization, International Programme on Chemical Safety, 55 pp (Concise International Chemical Assessment Document 24).

Isono RS, Kita J, & Setoguma T (1998) Acute effects of kaolinite suspension on eggs and larvae of some marine teleosts. Comp Biochem Physiol C, **120**: 449–455.

JCPDS (1981) Powder diffraction file search manual for common phases / inorganic and organic. Swarthmore, Pennsylvania, Joint Committee for Powder Diffraction Standards, Centre for Diffraction Data, 344 pp.

JECFA (1976) Evaluation of certain food additives and contaminants. Twentieth report of the Joint FAO/WHO Committee on Food Additives. Geneva, World Health Organization, p. 11 (WHO Technical Report Series).

Jepson CP (1984) Sodium bentonite: Still a viable solution for hazardous waste containment. Pollut Eng, **16**(7): 50–53.

Johnson NF, Haslam PL, Dewar A, Newman-Taylor AJ, & Turner-Warwick M (1986) Identification of inorganic dust particles in bronchoalveolar lavage macrophages by energy dispersive X-ray microanalysis. Arch Environ Health, **41**: 133–144.

Jones RA, Broder MW, & Stotzky G (1991) Effects of genetically engineered microorganisms on nitrogen transformations and nitrogen-transforming microbial populations in soil. Appl Environ Microbiol, **57**: 3212–3219.

Juhasz Z, Timár M, & Adamis Z (1978) Wirkung mechanischen Aktivierens auf Lebendzellen-Reaktionen mineralischer Stäube. Staub-Reinhalt Luft, **38**: 131–134.

Kalliomäki PL, Taikina-aho O, Paakko P, Anttila S, Kerola T, Sivonen SJ, Tienari J, & Sutinen S (1989) Smoking and the pulmonary mineral particle burden. IARC Sci Publ, **90**: 323–329.

Keatinge GF & Potter NM (1949) Health and environmental conditions in brickworkers. Br J Ind Med, **1**: 31–44.

Kennedy T, Rawlings W Jr, Baser M, & Tockman M (1983) Pneumoconiosis in Georgia kaolin workers. Am Rev Respir Dis, **127**: 215–220.

Kettle EH & Hilton R (1932) The technique of experimental pneumoconiosis. Lancet, **222**: 1190–1192.

King EJ, Harrison CV, & Nagelschmidt G (1948) The effects of kaolin on the lungs of rats. J Pathol Bacteriol, **60**: 435–440.

Kriegseis W, Scharmann A, & Serafin J (1987) Investigations of surface properties of silica dusts with regard to their cytotoxicity. Ann Occup Hyg, **31**: 417–427.

Kuzvart M (1984) Bentonite and montmorillonite clay. In: Industrial minerals and rocks. Elsevier, Amsterdam, pp 280–287 (Developments in Economic Geology 18).

Lapenas DJ & Gale PN (1983) Kaolin pneumoconiosis. A case report. Arch Pathol Lab Med, **107**: 650–653.

Lapenas D, Gale P, Kennedy T, Rawlings W Jr, & Dietrich P (1984) Kaolin pneumoconiosis. Radiologic, pathologic, and mineralogic findings. Am Rev Respir Dis, **130**: 282–288.

Le Bouffant L, Danile H, & Martin JC (1980) The values and limits of the relationship between cytotoxicity and fibrogenicity of various mineral dusts. In: Brown RC, Gormley IP, Chamberlain M, & Davies R ed. The *in vitro* effects of mineral dusts. London, Academic Press, pp 333–338.

Lee DR (1976) Development of an invertebrate bioassay to screen petroleum refinery effluents discharged into freshwater [PhD thesis]. Blacksburg, Virginia, Virginia Polytechnic and State University [cited in US EPA, 2004].

Lee RE (1993) Scanning electron microscopy and x-ray microanalysis. Englewood Cliffs, New Jersey, Prentice Hall, 458 pp.

Leiteritz H, Einbrodt HJ, & Martin JC (1967) Grain size and mineral content of lung dust of coal mines compared with mine dust. In: Davies CN ed. Inhaled particles and vapours. II. Oxford, Pergamon Press, pp 381–392.

Lemaistre P (1894) Congrès pour l'étude de la tuberculose, Paris, 1893, **3**: 491 [cited in King et al., 1948].

Lesser M, Zia M, & Kilburn KH (1978) Silicosis in kaolin workers and firebrick makers. South Med J, **71**: 1242–1246.

Levin JL, Frank AL, Williams MG, McConnel W, Suzuki Y, & Dodson R (1996) Kaolinosis in a cotton mill worker. Am J Ind Med, **29**: 215–221.

Lines M (2003) Asian ceramic clays. Ind Miner, **427**: 50–55.

Lipson SM & Stotzky G (1983) Adsorption of reovirus to clay minerals: Effects of cation-exchange capacity, cation saturation, and surface area. Appl Environ Microbiol, **46**: 673–682.

Love RG, Waclawski ER, Maclaren WM, Wetherill GZ, Groat SK, Porteous RH, & Soutar CA (1999) Risks of respiratory disease in the heavy clay industry. Occup Environ Med, **56**: 124–133.

Low RB, Leffingwell CM, & Bulman CA (1980) Effects of kaolinite on amino acid transport and incorporation into protein by rabbit pulmonary alveolar macrophages. Arch Environ Health, **35**: 217–223.

Lynch KM & McIver FA (1954) Pneumoconiosis from exposure to kaolin dust: kaolinosis. Am J Pathol, **30**: 1117–1127.

Ma L & Tang J (2002) Refined kaolin in China. Quality improvements needed to meet future paper demand. Ind Miner, **415**: 66–71.

Mackenzie RC ed. (1970) IV. The smectite group, A. Dioctahedral smectites. In: Differential thermal analysis. London, Academic Press, pp 442–452, 504–527.

Mányai S, Kabai J, Kis J, Süveges E, & Timár M (1969) The *in vitro* hemolytic effect of various clay minerals. Med Lav, **60**: 331–342.

Mányai S, Kabai J, Kis J, Suveges E, & Timár M (1970) The effect of heat treatment on the structure of kaolin and its *in vitro* hemolytic activity. Environ Res, **3**: 187–198.

Marek J (1981) Bentonite-induced rat paw oedema as a tool for simultaneous testing of prophylactic and therapeutic effects of anti-inflammatory and other drugs. Pharmazie, **36**: 46–49, 370.

Marek J & Blaha V (1982) Morphology of the bentonite and kaolin-induced rat-paw oedemas. Int J Tissue React, **4**: 103–114.

Martin JC, Danile H, & Le Bouffant L (1977) Short- and long-term experimental study of the toxicity of coal-mine dust and some of its constituents. In: Walton WH & McGovern B ed. Inhaled particles. IV. Oxford, Pergamon Press, pp 361–370.

Mascolo N, Summa V, & Tateo F (2004) *In vivo* experimental data on the mobility of hazardous chemical elements from clays. Appl Clay Sci, **25**: 23–28.

Mastin JP, Furbish WJ, De Long ER, Roggli VL, Pratt PC, & Shelburne JD (1986) In: Romig AD Jr & Chambers WF ed. Microbeam analysis 1986: 21st annual conference of the Microbeam Analysis Society, Albuquerque, New Mexico. San Francisco, California, San Francisco Press, pp 583–585.

McDonough S (1997) Bentonite toxicosis in a cat from ingestion of clay cat litter? Vet Hum Toxicol, **39**(3): 181–182.

McFarland VA & Peddicord RK (1980) Lethality of a suspended clay to a diverse selection of marine and estuarine macrofauna. Arch Environ Contam Toxicol, **9**: 733–741.

McLaren AD, Peterson GH, & Barshad I (1958) The adsorption and reactions of enzymes and proteins on clay minerals. IV. Kaolinite and montmorillonite. Soil Sci Soc Am Proc, **22**: 239–243.

McNally WD & Trostler IS (1941) Severe pneumoconiosis caused by inhalation of Fuller's earth. J Ind Hyg Toxicol, **23**: 118–126.

Melkonjan AM, Agekjan FG, & Muradjan SP (1981) [Problems of occupational hygiene in bentonite powder production.] Gig Tr Prof Zabol, **3**: 19–21 (in Russian).

Meredith TJ & Vale JA (1987) Treatment of paraquat poisoning in man: Methods to prevent absorption. Hum Toxicol, **6**: 49–55.

Middleton EL (1936) Industrial pulmonary disease due to the inhalation of dust, with special reference to silicosis. Lancet, **231**(ii): 59–64.

Middleton EL (1940) Silicosis. Proceedings of the international conference. Geneva, International Labour Office, pp 25, 134 (Studies and Reports, Series F [Industrial Hygiene], No. 17) [cited in McNally & Trostler, 1941].

Mikhailova-Docheva L, Buralkov T, & Kolev K (1986) [Hygienic evaluation of the occupational risk in working with Bulgarian bentonite raw material.] Probl Khig, **11**: 106–113 (in Bulgarian).

Miller RW, Honarvar S, & Hunsaker B (1980) Effects of drilling fluids on soils and plants: i. Individual fluid components. J Environ Qual, **9**: 547–552.

Monsó E, Tura JM, Pujadas J, Morell F, Ruiz J, & Morera J (1991) Lung dust content in idiopathic fibrosis: a study with scanning electron microscopy and energy dispersive x ray analysis. Br J Ind Med, **48**: 327–331.

Monsó E, Carreres A, Tura JM, Ruiz J, Fiz J, Xaus C, Llatjós M, & Morera J (1997) Electron microscopic microanalysis of bronchoalveolar lavage: a way to identify exposure to silica and silicate dust. Occup Environ Med, **54**: 560–565.

Moore DM & Reynolds RC Jr (1989) X-ray diffraction and the identification and analysis of clay minerals. Oxford, Oxford University Press, 332 pp.

Moore P (2003) Kaolin — white gold or white dirt? Ind Miner, **430**: 14–35.

Morgan WK, Donner A, Higgins ITT, Perason MG, & Rawlings W Jr (1988) The effects of kaolin on the lung. Am Rev Respir Dis, **138**: 813–820.

Mortensen JL (1961) Adsorption of hydrolyzed polyacrylonitrile on kaolinite. Clays Clay Miner, **9**: 530–545.

Mossman BT & Craighead JE (1982) Comparative cocarcinogenic effects of crocidolite asbestos, hematite, kaolin and carbon in implanted tracheal organ cultures. Ann Occup Hyg, **26**: 553–567.

Murphy EJ, Roberts E, Anderson DK, & Horrocks LA (1993a) Cytotoxicity of aluminium silicates in primary neuronal cultures. Neuroscience, **57**: 483–490.

Murphy EJ, Roberts E, & Horrocks LA (1993b) Aluminium silicate toxicity in cell cultures. Neuroscience, **55**: 597–605.

Murray HH (1994) Common clay. In: Carr DD ed. Industrial minerals and rocks. Littleton, Colorado, Society for Mining, Metallurgy, and Exploration, pp 247–248.

Myers JE & Cornell JE (1989) Respiratory health of brickworkers in Cape Town, South Africa. Symptoms, signs and pulmonary function abnormalities. Scand J Work Environ Health, **15**: 188–194.

Myers JE, Lewis P, & Hofmeyr W (1989a) Respiratory health of brickworkers in Cape Town, South Africa. Background, aims and dust exposure determinations. Scand J Work Environ Health, **15**: 180–187.

Myers JE, Garisch D, & Louw SJ (1989b) Respiratory health of brickworkers in Cape Town, South Africa. Radiographic abnormalities. Scand J Work Environ Health, **15**: 195–197.

Narang S, Rahman Q, Kaw JL, & Zaidi SH (1977) Dissolution of silicic acid from dusts of kaolin, mica and talc and its relation to their hemolytic activity — an *in vitro* study. Exp Pathol (Jena), **13**: 346–349.

Nemmar A, Hoylaerts MF, Hoet PH, & Nemery B (2004) Possible mechanisms of the cardio-vascular effects of inhaled particles: systemic translocation and prothrombic effects. Toxicol Lett, **149**: 243–253.

Newbury DE, Swyt CR, & Myklebust RL (1995) "Standardless" quantitative electron probe microanalysis with energy-dispersive x-ray spectrometry: is it worth the risk? Anal Chem, **67**: 1866–1871.

Oberson D, Hornebeck W, Pezerat H, Sebastien P, & Lafuma C (1993) Environmental mineral dust particles may be inhibitors of human leucocyte elastase. Possible involvement in occupational lung diseases. In: Hurych J, Lesage M, & David A ed. Proceedings of the 8th international conference on occupational lung diseases. Prague, Czech Medical Society, pp 1173–1179.

Oberson D, Desfontaines L, Pezerat H, Hornebeck W, Sebastien P, & Lafuma C (1996) Inhibition of human leukocyte elastase by mineral dust particles. Am J Physiol, **270**: L761–L771.

Ogle CJ, Rundle EM, & Sugar ET (1989) China clay workers in the south west of England: analysis of chest radiograph readings, ventilatory capacity, and respiratory symptoms in relation to type and duration of occupation. Br J Ind Med, **46**: 261–270.

Oldham PD (1983) Pneumoconiosis in Cornish china clay workers. Br J Ind Med, **40**: 131–137.

Orth H & Nisi D (1980) [Dust measurements in an iron foundry.] Giesserei, **67**: 62–67 (in German).

Oscarson DW, Van Scoyoc GE, & Ahlrichs JL (1981) Effect of poly-2-vinylpyridine-*N*-oxide and sucrose on silicate-induced hemolysis of erythrocytes. J Pharm Sci, **70**: 657–659.

Oscarson DW, Van Scoyoc GE, & Ahlrichs JL (1986) Lysis of erythrocytes by silicate minerals. Clays Clay Miner, **34**: 74–80.

Osornio-Vargas AR, Hernandez-Rodriguez NA, Yanez-Buruel AG, Ussler W, Overby LH, & Brody AR (1991) Lung cell toxicity experimentally induced by a mixed dust from Mexicali, Baja California, Mexico. Environ Res, **56**: 31–47.

Ottery J & Gormley IP (1978) Some factors affecting the haemolytic activity of silicate minerals. Ann Occup Hyg, **21**: 131–139.

Oudiz J, Brown JW, Ayer H, & Samuels S (1983) A report on silica exposure levels in United States foundries. Am Ind Hyg Assoc J, **44**: 374–376.

Oxman AD, Muir DC, Shannon HS, Stock SR, Hnizdo E, & Lange HJ (1993) Occupational dust exposure and chronic obstructive pulmonary disease. Am Rev Respir Dis, **148**: 38–48.

Pabst M & Hofer F (1998) Deposits of different origin in the lungs of the 5,300-year-old Tyrolean Iceman. Am J Phys Anthropol, **107**: 1–12.

Paoletti L, Batisti D, Caiazza S, Petrelli MG, Taggi F, De Zorzi L, Dina MA, & Donelli G (1987) Mineral particles in the lungs of subjects resident in the Rome area and not occupationally exposed to mineral dust. Environ Res, **44**: 18–28.

Parker SP ed. (1988) McGraw-Hill encyclopedia of the geological sciences, 2nd ed. New York, McGraw-Hill, pp 32–33, 69–72, 400–401.

Parkes WR (1982) Occupational lung disorders. London, Butterworths, pp 310–318.

Patterson EC & Staszak DJ (1977) Effects of geophagia (kaolin ingestion) on the maternal blood and embryonic development in the pregnant rat. J Nutr, **107**: 2020–2025.

Patterson SH & Murray HH (1975) Clays. In: Lefond SI ed. Industrial minerals and rocks, 4th ed. New York, American Institute of Mining, Metallurgical, and Petroleum Engineers, pp 519–595.

Patterson SH & Murray HH (1983) Clays. In: Lefond SI ed. Industrial minerals and rocks, 5th ed. New York, American Institute of Mining, Metallurgical, and Petroleum Engineers, pp 585–651.

Pekkanen J, Timonen KL, Ruuskanen J, Reponen A, & Mirme A (1997) Effects of ultrafine and fine particles in urban air on peak expiratory flow among children with asthmatic symptoms. Environ Res, **74**: 24–33.

Perry KMA (1947) Diseases of the lung resulting from occupational dusts other than silica. Thorax, **2**: 91–120.

Phibbs BP, Sundin RE, & Mitchell RS (1971) Silicosis in Wyoming bentonite workers. Am Rev Respir Dis, **103**: 1–17.

Policard A & Collet A (1954) [Experimental study on pathological effect of kaolin.] Schweiz Z Pathol Bakteriol, **17**: 320–325 (in German).

Potter EV & Stollerman GH (1961) The opsonization of bentonite particles by gamma-globulin. J Immunol, **87**: 110–118.

Que-Hee SS (1989) Respirable/total dust and silica content in personal air sample in a non-ferrous foundry. Appl Ind Hyg, **4**: 57–60.

Rajhans GS & Budlovsky J (1972) Dust conditions in brick plants of Ontario. Am Ind Hyg Assoc J, **33**: 258–268.

Rees D, Cronje R, & du Toir RSJ (1992) Dust exposure and pneumoconiosis in a South African pottery. 1. Study objectives and dust exposure. Br J Ind Med, **49**: 459–464.

Rice C, Harris RL, Lumsden JC, & Symons MJ (1984) Reconstruction of silica exposure in the North Carolina dusty trades. Am Ind Hyg Assoc J, **45**: 689–695.

Rieder M, Cavazzini G, D'yakonov YS, Frank-Kamenetskii VA, Gottardi G, Guggenheim S, Koval PV, Müller G, Neiva AMR, Radoslovich EW, Robert J-L, Sassi FP, Takeda H, Weiss Z, & Wones DR (1998) Nomenclature of micas. Clays Clay Miner, **46**: 586–595.

Righi F & Meunier A (1995) Origin of clays by rock weathering and soil formation. In: Velde B ed. Origin and mineralogy of clays. Berlin, Springer-Verlag, pp 103–161.

Robertson A, Dodgson J, Gormley IP, & Collings P (1982) An investigation of the adsorption of oxides of nitrogen on respirable mineral dusts and the effects on their cytotoxicity. Ann Occup Hyg, **26**: 607–624.

Rodeheaver G, Pettry D, Turnbull V, Edgerton MT, Edlich RF (1974) Identification of the wound infection-potentiating factors in soil. Am J Surg, **128**: 8–14.

Rombola G & Guardascione V (1955) La silicosi da bentoni. Med Lav, **46**: 480–497.

Rosmanith J, Hilscher W, Hessling B, Schyma SB, & Ehm W (1989) The fibrogenic action of kaolinite, muscovite and feldspar. In: Results of studies on dust suppression and silicosis prevention in hard coal mining, Vol 17. Essen, Steinkohlenbergbauverein, pp 305–321.

Roy R (1949) Decomposition and resynthesis of the micas. J Am Ceram Soc, **32**: 202–210.

RTECS (2003a) Bentonite. RTECS No. CT9450000. Atlanta, Georgia, US Department of Health and Human Services, Centers for Disease Control and Prevention, National Institute for Occupational Safety and Health, Registry of Toxic Effects of Chemical Substances. Available at: http://www.cdc.gov/niosh/rtecs/ct903210.html (accessed on 8 November 2004).

RTECS (2003b) Clay (kaolin). RTECS No. GF1670500. Atlanta, Georgia, US Department of Health and Human Services, Centers for Disease Control and Prevention, National Institute for Occupational Safety and Health, Registry of Toxic Effects of Chemical Substances. Available at: http://www.cdc.gov/niosh/rtecs/gf197d64.html#L (accessed on 8 November 2004).

Rundle EM, Sugar ET, & Ogle CJ (1993) Analyses of the 1990 chest health survey of china clay workers. Br J Ind Med, **50**: 913–919.

Rüttner JR, Bovet P, Weber R, & Willy W (1952) Neue Ergebnisse tierexperimenteller Silikoseforschung. Naturwissenschaften, **39**: 332.

Sadler RL & Roy TJ (1990) Smoking and mortality from coalworkers' pneumoconiosis. Br J Ind Med, **47**: 141–142.

Sahle W, Sallsten G, & Thoren K (1990) Characterization of airborne dust in a soft paper mill. Ann Occup Hyg, **34**: 55–75.

Sahu AP, Shanker R, & Zaidi SH (1978) Pulmonary response to kaolin, mica and talc in mice. Exp Pathol (Jena), **16**: 276–282.

Sakula A (1961) Pneumoconiosis due to Fuller's earth. Thorax, **16**: 176–179.

Salt PD (1985) Quantitative mineralogical analysis of small samples of china clay using x ray diffractometry. Br J Ind Med, **42**: 635–641.

Schiffenbauer M & Stotzky G (1982) Adsorption of coliphages T1 and T7 to clay minerals. Appl Environ Microbiol, **43**: 590–596.

Schindler PW & Stumm W (1987) The surface chemistry of oxides, hydroxides, and oxide minerals. In: Stumm W ed. Aquatic surface chemistry: chemical processes at the particle–water interface. New York, Wiley-Interscience, pp 83–110.

Schmidt KG & Lüchtrath H (1958) [Effect of fresh and burned kaolin on the lungs and peritoneum of rats.] Beitr Silikoseforsch, **119**: 3–37 (in German).

Schreider JP, Culbertson MR, & Raabe OG (1985) Comparative pulmonary fibrogenic potential of selected particles. Environ Res, **38**: 256–274.

Seaton A, Lamb D, Brown WR, Sclare G, & Middleton WG (1981) Pneumoconiosis of shale miners. Thorax, **36**: 412–418.

Sébastien P, Chamak B, Gaudichet A, Bernaudin JF, Pinchon MC, & Bignon J (1994) Comparative study by analytical transmission electron microscopy of particles in alveolar and interstitial human lung macrophages. Ann Occup Hyg, **38**(Suppl 1): 243–250.

Sepulveda MJ, Vallyathan V, Attfield MD, Piacitelli L, & Tucker JH (1983) Pneumoconiosis and lung function in a group of kaolin workers. Am Rev Respir Dis, **127**: 231–235.

Sheers G (1964) Prevalence of pneumoconiosis in Cornish kaolin workers. Br J Ind Med, **21**: 218–225.

Smykatz-Kloss W (1974) Differential thermal analysis. New York, Springer-Verlag, pp 1–23, 64–72, 77–91.

Snipes MB, Boecker BB, & McClellan RO (1983a) Retention of monodisperse or polydisperse aluminosilicate particles inhaled by dogs, rats, and mice. Toxicol Appl Pharmacol, **69**: 345–362.

Snipes MB, Muggenburg BA, & Bice DE (1983b) Translocation of particles from lung lobes or the peritoneal cavity to regional lymph nodes in beagle dogs. J Toxicol Environ Health, **11**: 703–712.

Sprague JB & Logan WJ (1979) Separate and joint toxicity to rainbow trout of substances used in drilling fluids for oil exploration. Environ Pollut, **19**: 269–281.

Srodon J & Eberl DD (1984) Illite. In: Bailey SW ed. Micas. Blacksburg, Virginia, Mineralogical Society of America, pp 495–554 (Reviews in Mineralogy, Vol 13).

Starkey HC, Blackmon PD, & Hauff PL (1984) The routine mineralogical analysis of clay-bearing samples. Washington, DC, US Geological Survey, 32 pp (US Geological Survey Bulletin 1563).

Steel RF & Anderson W (1972) The interaction between kaolinite and *Staphylococcus aureus*. J Pharm Pharmacol, **24**(Suppl): 1–129.

Stobbe TJ, Plummer RW, Kim H, & Dower JM (1986) Characterisation of coal mine dust. Ann Am Conf Gov Ind Hyg, **14**: 689–696.

Stölzel M, Peters A, & Wichmann H-E (2003) Daily mortality and fine and ultrafine particles in Erfurt, Germany. In: Revised analyses of time-series studies of air pollution and health. Special report. Boston, Massachusetts, Health Effects Institute, pp 231–240. Available at: http://www.healtheffects.org/news.htm (accessed on 16 May 2003).

Stumm W (1997) Reactivity at the mineral–water interface: dissolution and inhibition. Colloids Surf A, **120**: 143–166.

Stumm W & Wollast R (1990) Coordination chemistry of weathering: kinetics of surface-controlled dissolution of oxide minerals. Rev Geophys, **28**: 53–69.

Styles JA & Wilson J (1973) Comparison between *in vitro* toxicity of polymer and mineral dusts and their fibrogenicity. Ann Occup Hyg, **16**: 241–250.

Sundius N, Bygdén A, & Bruce T (1936) Der Staubinhalt einer silikotischen Lunge eines Steingutarbeiters. Ber Dtsch Keram Ges, **17**: 73–90.

Sykes SE, Morgan A, Evans JC, Evans N, Holmes A, & Moores SR (1982) Use of an *in vivo* test system to investigate the acute and sub-acute responses of the rat lung to mineral dusts. Ann Occup Hyg, **26**: 593–605.

Tamás F (1982) Szilikátipari kézikönyv. Budapest, Müszaki Könyvkiadó.

Tátrai E, Adamis Z, Timár M, & Ungváry G (1983) Comparative histopathological and biochemical analysis of early stages of exposure to non-silicogenic aluminium silicate and strongly silicogenic quartz dusts in rats. Exp Pathol, **23**: 163–171.

Tátrai E, Ungváry G, Adamis Z, & Timár M (1985) Short-term *in vivo* method for prediction of the fibrogenic effect of different mineral dusts. Exp Pathol, **28**: 111–118.

Terashima T, Wiggs B, English D, Hogg JC, & Van Eeden SF (1997) Phagocytosis of small carbon particles (PM_{10}) by alveolar macrophages stimulates the release of polymorphonuclear leucocytes from bone marrow. Am J Respir Crit Care Med, **155**: 1441–1447.

Thomas RW (1952) Silicosis in the ball-clay and china-clay industries. Lancet, **1**(3): 133–135.

Thorez J (1975) Phyllosilicates and clay minerals: a laboratory handbook for their x-ray diffraction analysis. Dison, Editions G. Lelotte, 579 pp.

Timár M, Kendrey G, & Juhasz Z (1966) Experimental observations concerning the effects of mineral dust to pulmonary tissue. Med Lav, **57**: 1–9.

Timár M, Adamis Z, & Ungváry G (1979) Biological effects of mineral dusts. *In vitro* and *in vivo* studies. Arch Hig Rada Toksikol, **30**(Suppl): 871–874.

Timár M, Adamis Z, Tátrai E, & Ungváry G (1980) *In vivo* and *in vitro* investigations on different dusts. In: Brown RC, Gormley IP, Chamberlain M, & Davies R ed. The *in vitro* effects of mineral dusts. London, Academic Press, pp 319–322.

Tonning HO (1949) Pneumoconiosis from Fuller's earth. J Ind Hyg Toxicol, **31**: 41–45.

Ungváry G, Timár M, Tátrai E, Bacsy E, & Gaal G (1983) Analysis of aluminium silicate storage foci in the lungs. Exp Pathol, **23**: 203–214.

Unsworth EF, Pearce J, McMurray CH, Moss BW, Gordon FJ, & Rice D (1989) Investigations of the use of clay minerals and Prussian Blue in reducing the transfer of dietary radiocesium to milk. Sci Total Environ, **85**: 339–347.

Uragoda CG & Fernando BND (1974) An investigation into the health of kaolin workers in Sri Lanka. Ceylon Med J, **19**: 77–79.

US Department of Labor (1990) Health inspection procedures. Washington, DC, US Department of Labor, Mine Safety and Health Administration, pp C-1 – E-14 (Handbook No. PH90-IV-4).

US EPA (1996a) Code of Federal Regulations, Part 40: 50.6 (p 652), Appendix I (pp 722–730), and 53.40 (pp 41–49). Available at: http://a257.g.akamaitech.net/7/257/2422/12feb20041500/edocket.access.gpo.gov/cfr_2004/julqtr/pdf/40cfr50.6.pdf and http://a257.g.akamaitech.net/7/257/2422/12feb20041500/edocket.access.gpo.gov/cfr_2004/julqtr/pdf/40cfr53.4.pdf

US EPA (1996b) Ambient levels and noncancer health effects of inhaled crystalline and amorphous silica: health issue assessment. Washington, DC, US Environmental Protection Agency, Office of Research and Development (Publication No. EPA/600/R-95/115; National Technical Information Service Publication No. PB97-188122).

US EPA (2004) ECOTOX database. Washington, DC, US Environmental Protection Agency. Available at: http://www.epa.gov/ecotox/.

US FDA (2004) Code of Federal Regulations. Title 1. Food and Drugs. Part 84. Direct food substances affirmed as generally recognized as safe (GRAS). Washington, DC, US Department of Health and Human Services, Food and Drug Administration. Available at: http://www.access.gpo.gov/nara/cfr/waisidx_98/21cfr184_98.html.

USGS (1997) U.S. Department of Interior & U.S. Geological Survey mineral industry surveys: Clay and shale annual review 1995. Reston, Virginia, US Department of Interior, 8 pp + appendices.

USGS (2001) A laboratory manual for x-ray powder diffraction. Washington, DC, US Geological Survey (Open-file report 01-041). Available at: http://pubs.usgs.gov/of/of01-041/htmldocs/clays/illite.htm (last modified on 11 October 2001).

Vallyathan V, Sepulveda MJ, Tucker JG, Green FHY (1982) Kaolinite pneumoconiosis in Cornish kaolin workers. Am Rev Respir Dis, **127**: 231–235.

Vallyathan V, Schwegler D, Reasor M, Stettler L, Clere J, & Green FHY (1988) Comparative *in vitro* cytotoxicity and relative pathogenicity of mineral dusts. Ann Occup Hyg, **32**(Suppl.): 279–289.

Van der Marel HW & Beutelspacher H (1976) Atlas of infrared spectroscopy of clay minerals and their mixtures. Oxford, Elsevier, 396 pp.

Velde B (1985) Clay minerals. A physicochemical explanation of their occurrence. Amsterdam, Elsevier, pp 70–82.

Velde B ed. (1995) Composition and mineralogy of clay minerals. In: Origin and mineralogy of clays. Berlin, Springer-Verlag, pp 27–33 (Clays and the Environment, Vol 1).

Virta RL (1997) US Geological Survey mineral industry surveys. Clays — 1997. Available at: http://minerals.usgs.gov/minerals/pubs/commodity/clays/190497.pdf.

Virta RL (1999) US Geological Survey mineral industry surveys. Clay and shale — 1999. Available at: http://minerals.usgs.gov/minerals/pubs/commodity/clays/190499.pdf.

Virta RL (2001) US Geological Survey mineral industry surveys. Clay and shale — 2001. Available at: http://minerals.usgs.gov/minerals/pubs/commodity/clays/claymyb01.pdf.

Virta RL (2002) US Geological Survey mineral industry surveys. Clay and shale — 2002. Available at: http://minerals.usgs.gov/minerals/pubs/commodity/clays/claysmyb02.pdf.

Wagner JC, Pooley FD, Gibbs A, Lyons L, Sheers G, & Moncrieff CB (1986) Inhalation of china stone and clay dust: Relationship between the mineralogy of dust retained in the lungs and pathological changes. Thorax, **41**: 190–196.

Wagner JC, Griffiths DM, & Munday DE (1987) Experimental studies with palygorskite dusts. Br J Ind Med, **44**: 749–763.

Wallace WE, Headley LC, & Weber KC (1975) Dipalmitoyl lecithin surfactant adsorption by kaolin dust *in vitro*. J Colloid Interface Sci, **51**: 535–537.

Wallace WE Jr, Vallyathan V, Keane MJ, & Robinson V (1985) *In vitro* biologic toxicity of native and surface-modified silica and kaolin. J Toxicol Environ Health, 16: 415–424.

Walsh PN (1972) The effects of collagen and kaolin on the intrinsic coagulant activity of platelets. Evidence for an alternative pathway in intrinsic coagulation not requiring factor XII. Br J Haematol, **22**: 393–405.

Warraki S & Herant Y (1963) Pneumoconiosis in china-clay workers. Br J Ind Med, **20**: 226–230.

Weber JB (1970) Adsorption of s-triazines by montmorillonite as a function of pH and molecular structure. Soil Sci Soc Am Proc, **34**: 401–404.

Weber JB, Perry PW, & Upchurch RP (1965) The influence of temperature and time on the adsorption of paraquat, diquat, 2,4-D and prometone by clays, charcoal, and an anion-exchange resin. Soil Sci Soc Am Proc, **29**: 678–688.

White R & Kuhn C (1980) Effects of phagocytosis of mineral dusts on elastase secretion by alveolar and peritoneal exudative macrophages. Arch Environ Health, **35**: 106–109.

Wiecek E, Goscicki J, Indulski J, & Stroszejn-Mrowea G (1983) [Dust and occupational diseases in brick yards.] Med Pr, **34**: 34–45 (in Polish).

Wiles M, Huebner H, Afriyie-Gyawu E, Taylor R, Bratton G, & Phillips T (2004) Toxicological evaluation and metal bioavailability in pregnant rats following exposure to clay minerals in the diet. J Toxicol Environ Health A, **67**(11): 863–874.

Wilken RD & Wirth H (1986) The adsorption of hexachlorobenzene on naturally occurring adsorbents in water. IARC Sci Publ, **77**: 75–81.

Wilson JW (1951) Hepatomas in mice fed a synthetic diet low in protein and deficient in choline. Cancer Res, **11**: 290 (abstr).

Wilson JW (1953a) Nutritional deficiency produced in the mouse by feeding bentonite. J Natl Cancer Inst, **14**: 57–63.

Wilson JW (1953b) Hepatomas in mice on a diet of bentonite. J Natl Cancer Inst, **14**: 65–71.

Wilson JW (1954) Hepatomas produced in mice by feeding bentonite in the diet. Ann N Y Acad Sci, **57**: 678–686.

Wojtczak J, Bielichowska G, Stroszejn-Mrowca G, & Tenerowicz B (1989) [Fly ash and its biological effects. 3. Exposure to dust of workers in the energy-generating industry (power plants and thermoelectric power stations).] Med Pr, **40**: 294–301.

Woodworth CD, Mossman BT, & Craighead JE (1982) Comparative effects of fibrous and nonfibrous minerals on cells and liposomes. Environ Res, **27**: 190–205.

Wynder EL & Hoffman D (1967) Tobacco and tobacco smoke: Studies in experimental carcinogenesis. New York, Academic Press, 730 pp.

Yamada H, Hashimoto H, Akiyama M, Kawabata Y, & Iwai K (1997) Talc and amosite/crocidolite preferentially deposited in the lungs of nonoccupational female lung cancer cases in urban areas of Japan. Environ Health Perspect, **105**: 504–508.

Zaidi SH, Dogra RKS, Khanna S, & Shanker R (1981) Experimental infective pneumoconiosis: Effect of fibrous and non-fibrous silicates and *Candida albicans* on the lungs of guinea pigs. Ind Health, **19**: 85–91.

RESUME

1. Identité, propriétés physiques et chimiques et méthodes d'analyse

La bentonite est une roche composée d'argiles fortement colloïdales et plastiques, principalement de la montmorillonite, un minéral argileux du groupe de la smectite. Elle est produite par dévitrification *in situ* de cendre volcanique. Outre la montmorillonite, la bentonite peut renfermer du feldspath, de la cristobalite et du quartz cristallin. Entre autres propriétés spécifiques, la bentonite présente une capacité à former des gels thixotropes avec l'eau et à absorber de grandes quantités d'eau, ainsi qu'une forte capacité d'échange de cations. Les propriétés de la bentonite découlent de la structure cristalline des minéraux du groupe de la smectite, qui se présente sous la forme d'un feuillet d'octaèdres d'alumine, placé entre deux feuillets de tétraèdres de silice. La quantité d'eau interstitielle et la nature des cations échangeables présents dans l'espace intercouches influent sur les propriétés de la bentonite et ainsi sur les usages industriels des différents types de bentonite. Par extension, le terme de bentonite s'applique dans l'industrie à toute argile présentant des propriétés similaires. La terre de Fuller est souvent une bentonite.

Le kaolin ou argile de Chine est un mélange de différents minéraux. Son principal composant est la kaolinite. Il contient souvent en outre du quartz, du mica, du feldspath, de l'illite et de la montmorillonite. La kaolinite est constituée de minuscules feuillets de cristaux tricliniques, dotés d'une morphologie pseudohexagonale. Elle se forme par altération météorique des roches. Elle possède une certaine capacité d'échange de cations.

Les principaux composants de l'argile commune et de l'argile litée sont l'illite et la chlorite. L'illite est aussi un composant des argiles plastiques pour céramiques. Il existe une ressemblance étroite entre l'illite et les micas, mais dans la première, le nombre d'atomes d'aluminium remplacés par des atomes de silicium et/ou le remplacement partiel des ions potassium entre les couches unitaires par d'autres cations, tels que l'hydrogène, le magnésium ou le calcium, sont moindres.

Les mesures quantitatives portant sur les poussières en suspension dans l'air contenant des aluminosilicates utilisent le plus souvent des méthodes gravimétriques. Pour identifier et quantifier les aluminosilicates, on recourt notamment à la diffraction de rayons X, à la microscopie électronique, à l'analyse par rayons X à dispersion d'énergie, à l'analyse thermique différentielle, à la diffraction d'électrons et à la spectroscopie infrarouge.

2. Sources d'exposition humaine et environnementale

La montmorillonite est partout présente à faible concentration dans le sol, la charge sédimentaire des eaux naturelles et les poussières en suspension dans l'air. Sa biodégradation et sa bioaccumulation dans la chaîne alimentaire semblent minimales, à supposer même qu'elles se produisent, et la dégradation abiotique de la bentonite présente dans d'autres minéraux s'effectue seulement à l'échelle des temps géologiques.

Parmi les principaux usages de la bentonite, on peut mentionner : la liaison du sable de fonderie dans les moules, l'absorption des graisses, des huiles et des déchets animaux, la préparation des boulettes de taconite (minerai de fer) et l'amélioration des propriétés des boues de forage. Les usages spéciaux comprennent l'utilisation comme ingrédient dans les céramiques, l'imperméabilisation et l'étanchéité dans les projets de génie civil tels que les sites de décharge et les dépôts de déchets radioactifs, l'utilisation comme charge, stabilisant ou diluant dans les adhésifs, les peintures, les cosmétiques et les médicaments, comme vecteurs dans les pesticides et les engrais et comme liant dans les aliments pour animaux, la clarification du vin et des huiles végétale, ainsi que la purification des eaux usées. Les usages de la terre de Fuller de type montmorillonite recoupent ceux de la bentonite.

L'utilisation du kaolin remonte au troisième siècle avant Jésus-Christ en Chine. Aujourd'hui, le kaolin est extrait des mines et employé en quantités importantes en vue d'un grand nombre d'usages industriels. Son application la plus importante est la production du papier, dans laquelle il sert de matériau d'enduction. On l'utilise en outre en grandes quantités dans les secteurs du caoutchouc, des matières plastiques, des céramiques, des produits chimiques, des produits pharmaceutiques et des cosmétiques.

On utilise principalement l'argile commune et l'argile litée, dont l'illite est souvent un composant majeur, dans la fabrication de briques extrudées et autres, de ciments portland et autres, de blocs de béton et de béton de structure et de produits réfractaires. Les revêtements routiers, les tuiles en céramique, ainsi que les céramiques et les verres, constituent d'autres usages importants.

3. Concentrations dans l'environnement et exposition humaine

Compte tenu de la diffusion étendue de la bentonite dans la nature et de son utilisation dans une très grande variété de produits de consommation, la population générale est exposée partout à de faibles concentrations de ce matériau.

On dispose de données limitées sur l'exposition professionnelle à la poussière de bentonite dans les mines, les usines de traitement et les industries utilisatrices. Les valeurs maximales rapportées pour les concentrations de poussières totales et inhalables sont respectivement de 1430 mg/m^3 et de 34,9 mg/m^3, bien que la plupart des valeurs soient inférieures à 10 mg/m^3 pour les poussières totales et à 5 mg/m^3 pour les poussières inhalables.

Le kaolin est un composant naturel du sol et se rencontre souvent dans l'air ambiant. L'extraction minière et le raffinage du kaolin impliquent une exposition considérable et on s'attend également à ce que la production du papier, du caoutchouc et des matières plastiques s'accompagne d'une exposition importante. On ne dispose d'informations quantitatives sur l'exposition professionnelle que pour un petit nombre de pays et d'industries. Les concentrations de poussières inhalables dans l'extraction minière et le traitement du kaolin sont habituellement inférieures à 5 mg/m^3.

4. Cinétique et métabolisme chez les animaux de laboratoires et les êtres humains

On ne dispose d'aucune donnée sur la cinétique ou le métabolisme de la montmorillonite, de la kaolinite ou de l'illite dans les conditions où l'on rencontre ces matériaux dans la plupart des milieux de travail.

On a étudié le dépôt et la cinétique de montmorillonite fondue radiomarquée chez la souris, le rat, le chien et l'homme après une exposition par inhalation. Le dépôt dans le nasopharynx augmente avec la taille de particule et prend une ampleur moindre chez le chien que chez les rongeurs. Le dépôt trachéobronchique est faible et indépendant de l'espèce animale et de la taille de particule. Le dépôt pulmonaire est considérablement plus important chez le chien que chez les rongeurs et diminue lorsque la taille de particule augmente.

L'élimination des particules des poumons s'effectue par solubilisation *in situ* et par clairance physique. Chez le chien, le principal mécanisme de clairance est la solubilisation. Chez les rongeurs, le premier mécanisme d'élimination est le transport physique. La clairance par élimination mécanique est lente, en particulier chez le chien. La demi-vie est initialement de 140 jours et passe à 6900 jours au 200e jour après l'exposition.

Chez l'homme, on observe, au bout de 6 jours, une rapide clairance initiale à partir de la région pulmonaire de 8 % pour les particules d'aluminosilicate ayant un diamètre aérodynamique de 1,9 µm et de 40 % pour celles dont le diamètre aérodynamique est de 6,1 µm. Puis, 4 % et 11 % respectivement des particules de l'une et l'autre tailles sont éliminées après une demi-vie de 20 jours, les particules restantes étant évacuées avec des demi-vies de 330 jours et de 420 jours respectivement.

Une forte proportion des particules ultrafines (< 100 nm) se dépose dans la région nasale. Ces particules peuvent franchir la barrière alvéolaire/capillaire.

5. **Effets sur les mammifères de laboratoire et dans les systèmes de test *in vitro***

La toxicité des argiles est largement déterminée par leur teneur en quartz. La présence de quartz dans les argiles étudiées s'oppose à l'établissement d'une estimation indépendante fiable de la fibrogénicité des autres composants de ces argiles.

Chez les rongeurs, une injection intratrachéale unique de bentonite et de montmorillonite présentant une faible teneur en quartz a déclenché des effets cytotoxiques dépendants de la dose et de la taille de particule, ainsi qu'une inflammation locale transitoire, se manifestant notamment par des oedèmes et donc par une augmentation du poids des poumons. Chez le rat, des expositions intratrachéales uniques à la bentonite ont donné lieu à la constitution de foyers de stockage dans les poumons 3 à 12 mois plus tard. Après l'exposition intratrachéale de rats à de la bentonite à forte teneur en quartz, on a également observé des fibroses. La bentonite accroît la sensibilité des souris aux infections pulmonaires.

On dispose de données limitées sur les effets d'expositions multiples à la montmorillonite ou à la bentonite sur les animaux de laboratoire. Des souris maintenues à un régime alimentaire contenant 10 ou 25 % de bentonite, mais par ailleurs conçu pour soutenir une croissance normale, ont présenté des vitesses de croissance légèrement réduites, tandis que pour d'autres souris recevant un régime similaire, mais renfermant 50 % de bentonite, la croissance était minimale, avec apparition d'une stéatose hépatique et finalement d'une fibrose du foie et d'hépatomes bénins (voir ci-après).

Les études *in vitro* des effets de la bentonite sur divers types de cellules de mammifère indiquent habituellement une forte cytotoxicité. Des concentrations de bentonite inférieures à 1,0 mg/ml et des particules de montmorillonite de diamètre inférieur à 5 µm provoquent une détérioration de la membrane et même une lyse cellulaire, ainsi des modifications fonctionnelles dans plusieurs types de cellules. La vitesse et le degré de lyse des érythrocytes ovins se révèlent dose-dépendants.

L'instillation intratrachéale de kaolin donne des foyers de stockage, une réaction à un corps étranger et une réaction exsudative diffuse. Après l'administration de doses élevées de kaolin (contenant 8 à 65 % de quartz), certaines études ont décrit l'apparition de fibroses, tandis que pour des doses plus faibles de kaolin, les quelques études disponibles n'ont pas relevé de fibrose.

Les données sur la toxicité de l'illite sont très limitées et inexistantes sur celle des autres composants des autres argiles importantes pour l'industrie. L'instillation intratrachéale d'illite ayant une teneur en quartz inconnue a induit une protéinose alvéolaire, augmenté le poids des poumons et provoqué la synthèse de collagène. L'illite présente une cytotoxicité limitée à l'égard des macrophages péritonéaux et une action hémolytique *in vitro*.

On ne dispose d'aucune étude satisfaisante sur la cancérogénicité de la bentonite. Dans le cadre d'une étude par inhalation et dans celui d'une étude avec injection intrapleurale, le kaolin n'a pas induit de tumeurs chez le rat. On ne dispose d'aucune étude sur la génotoxicité des argiles.

Des études très limitées et isolées n'ont pas mis en évidence de toxicité pour le développement chez le rat après une exposition orale à la bentonite ou au kaolin.

6.   Effets sur l'homme

La population générale est partout exposée à de faibles concentrations de montmorillonite et de kaolinite, les principaux composants respectivement de la bentonite et du kaolin, et d'autres minéraux argileux. Il n'existe pas de données sur les effets potentiels de telles expositions de faible intensité.

Les expositions professionnelles à long terme à la poussière de bentonite peuvent provoquer des lésions structurales et fonctionnelles des poumons. Cependant, les données disponibles sont insuffisantes pour établir incontestablement une relation dose-réponse ou même une relation de cause à effet en raison du manque d'informations sur la période sur laquelle s'est déroulée l'exposition et sur l'intensité de cette dernière, et de l'existence de facteurs de confusion, tels que l'exposition à la silice et à la fumée de tabac.

L'exposition à long terme au kaolin entraîne le développement de pneumoconioses diagnostiquées radiologiquement en relation avec l'exposition. On n'a signalé des détériorations nettes de la fonction respiratoire et des symptômes associés que dans les cas présentant des anomalies radiologiques importantes. La composition de l'argile, à savoir la quantité et la nature des minéraux qu'elle contient en dehors de la kaolinite, est un déterminant majeur des effets.

La bentonite, le kaolin et autres argiles contiennent souvent du quartz et il existe une relation causale entre l'exposition au quartz et la silicose et le cancer du poumon. On a rapporté des augmentations statistiquement significatives de l'incidence de la bronchite chronique et de l'emphysème pulmonaire, ou de la mortalité due à ces maladies après des expositions au quartz.

7. Effets sur d'autres organismes en laboratoire et sur le terrain

La bentonite et le kaolin présentent une faible toxicité pour les espèces aquatiques, dont un grand nombre ont été soumis à des essais.

8. Evaluation des risques pour l'homme et des effets sur l'environnement

D'après les données limitées issues des études portant sur des personnes exposées à la bentonite, la montmorillonite retenue ne semble provoquer que des modifications tissulaires non spécifiques sans gravité, similaires à celles décrites dans le spectre de modifications induites par la « maladie des petites voies aériennes causée par des poussières minérales » (accumulations péribronchiolaires et nodulaires de poussières contenant un matériau réfractaire [montmorillonite], associées à une fibrose interstitielle limitée). Certaines études rapportent des anomalies radiologiques.

Aucun cas de réaction fibrotique diffuse ou nodulaire marquée des tissus pulmonaires à la montmorillonite en l'absence de silice libre n'a été signalé. On ne peut déduire aucune estimation quantitative de la capacité de la bentonite à provoquer des effets pulmonaires indésirables.

L'exposition à long terme au kaolin peut entraîner une pneumoconiose relativement bénigne, connue sous le nom de kaolinose. On n'a observé de détérioration de la fonction pulmonaire que dans les cas présentant des modifications radiologiques notables. D'après les données relatives aux travailleurs en contact avec l'argile de chine au Royaume Uni, on peut estimer très grossièrement que le pouvoir toxique du kaolin est inférieur d'un ordre de grandeur au moins à celui du quartz.

La bentonite, le kaolin et d'autres argiles contiennent du quartz, dont on sait qu'il provoque des silicoses et des cancers du poumon.

On n'a identifié aucun signalement d'effets indésirables, locaux ou systémiques, résultant de l'usage étendu de la bentonite ou du kaolin dans les produits cosmétiques.

Les effets biologiques des minéraux argileux dépendent de leur composition minérale et de leur taille de particule. Le classement décroissant des capacités du quartz, de la kaolinite et de la montmorillonite à provoquer des lésions pulmonaires est cohérent avec ce que l'on sait des surfaces spécifiques actives de ces matériaux et de leur chimie de surface.

La bentonite et le kaolin présentent une faible toxicité à l'égard des organismes aquatiques.

RESUMEN

1. Identidad, propiedades físicas y químicas y métodos analíticos

La bentonita es una roca formada por arcillas muy coloidales y plásticas cuyo componente fundamental es la montmorillonita, mineral arcilloso del grupo de la esmectita, que se produce por la desvitrificación *in situ* de la ceniza volcánica. Además de la montmorillonita, la bentonita puede contener feldespato, cristobalita y cuarzo cristalino. Entre las propiedades especiales de la bentonita cabe mencionar su capacidad para formar geles tixotrópicos con el agua y para absorber grandes cantidades de ésta y una elevada capacidad de intercambio de cationes. Las propiedades de la bentonita se derivan de la estructura cristalina del grupo de la esmectita, consistente en una lámina de alúmina octaédrica situada entre otras dos de silicio tetraédrico. Las variaciones del agua intersticial y los cationes intercambiables del espacio interlaminar influyen en las propiedades de la bentonita y, por consiguiente, en las aplicaciones comerciales de sus distintos tipos. Por extensión, el término bentonita se aplica comercialmente a cualquier arcilla con propiedades semejantes. La tierra de fuller es con frecuencia una bentonita.

El caolín o arcilla caolínica es una mezcla de minerales diferentes. Su principal componente es la caolinita; con frecuencia contiene además cuarzo, mica, feldespato, ilita y montmorillonita. La caolinita está formada por capas diminutas de cristales triclínicos de morfología pseudohexagonal. Se forma por la meteorización de las rocas y tiene cierta capacidad de intercambio de cationes.

Los principales componentes de la arcilla común y la pizarra son la ilita y la clorita. La ilita es también un componente de las arcillas de alfarero. Es muy parecida a las micas, pero tiene un grado de sustitución menor de silicio por aluminio y/o una sustitución parcial de iones de potasio entre las capas unitarias por otros cationes, como hidrógeno, magnesio y calcio.

La medición cuantitativa del polvo suspendido en el aire que contiene silicatos de aluminio se suele realizar normalmente mediante análisis gravimétrico. Los métodos utilizados para la identificación y cuantificación de los silicatos de aluminio son la difracción por rayos X, la microscopía electrónica, el análisis de energía dispersiva de rayos X, el análisis térmico diferencial, la difracción electrónica y la espectroscopia de infrarrojos.

2. Fuentes de exposición humana y ambiental

La montmorillonita suele estar presente en concentraciones bajas en el suelo, en los sedimentos de las aguas naturales y en el polvo suspendido en el aire. La biodegradación y la bioacumulación en la cadena alimentaria parecen ser mínimas, o nulas, y la degradación abiótica de la bentonita en otros minerales sólo se produce a una escala de tiempo geológica.

Las principales aplicaciones de la bentonita son la aglutinación de la arena de fundición en los moldes; la absorción de grasa, aceite y desechos de animales; la nodulización del mineral de hierro de la taconita; y la mejora de las propiedades de los lodos de perforación. Entre sus aplicaciones especiales cabe mencionar su empleo como ingrediente de la cerámica; en la impermeabilización y el sellado en proyectos de ingeniería civil, por ejemplo de vertederos y depósitos de desechos nucleares; como relleno, estabilizador o diluyente de adhesivos, pinturas, cosméticos y medicinas, como excipiente en plaguicidas y fertilizantes y como aglutinante en los piensos; como clarificador del vino y el aceite vegetal; y como purificador de aguas residuales. Las aplicaciones de la tierra de fuller del tipo de la montmorillonita se superponen con las de la bentonita.

El uso del caolín se remonta al siglo III a.C. en China. En la actualidad se extrae y utiliza en cantidades significativas para numerosos usos industriales. Su aplicación más importante es la producción de papel como material de revestimiento. Además, se usa en grandes cantidades en las industrias de la pintura, el caucho, el plástico, la cerámica, los productos químicos, los productos farmacéuticos y la cosmética.

La arcilla y el esquisto bituminoso, de los que a menudo la ilita es uno de los componentes principales, se utilizan sobre todo en la fabricación ladrillos (por extrusión o de otros tipos), cemento portland o de otros tipos, bloques y estructuras de hormigón, y material refractario. Encuentran también aplicaciones importantes en el revestimiento de carreteras, las baldosas cerámicas y en cerámicas y vidrios.

3. Niveles ambientales y exposición humana

Habida cuenta de la amplia distribución de la bentonita en la naturaleza y de su utilización en una enorme variedad de productos de consumo, la exposición de la población general a bajas concentraciones es ubicua.

La información relativa a la exposición ocupacional al polvo de bentonita en las minas, las instalaciones de elaboración y las industrias que la utilizan es limitada. Los valores máximos notificados para las concentraciones de polvo total y polvo respirable fueron, respectivamente, de 1430 y 34,9 mg/m^3, aunque la mayor parte fueron inferiores a 10 mg/m^3 para el polvo total y a 5 mg/m^3 para el polvo respirable.

El caolín es un componente natural del suelo y se encuentra ampliamente distribuido en el aire del medio ambiente. Su extracción y refinado conlleva una exposición considerable, que también es previsible en la producción de papel, caucho y plástico. Sólo se dispone de información cuantitativa sobre la exposición ocupacional para un pequeño número de países y de industrias. Las concentraciones de polvo respirable en la extracción y elaboración del caolín suelen ser inferiores a 5 mg/m^3.

4. Cinética y metabolismo en animales de laboratorio y en el ser humano

No se obtuvo información sobre la cinética o el metabolismo de la montmorillonita, la caolinita o la ilita tal como se encuentran en la mayoría de los lugares de trabajo.

Se ha estudiado la deposición y la cinética de la montmorillonita fundida radiomarcada en ratones, ratas, perros y personas tras su exposición por inhalación. La deposición nasofaríngea aumenta con el tamaño de las partículas y es menor en los perros que en los roedores. La deposición traqueobronquial fue escasa e independiente de la especie animal y del tamaño de las partículas. La deposición pulmonar fue considerablemente más alta en los perros que en los roedores y disminuía al aumentar el tamaño de las partículas.

La eliminación de las partículas de los pulmones tuvo lugar mediante solubilización *in situ* y arrastre físico. En los perros, el mecanismo principal de eliminación fue la solubilización y en los roedores el transporte físico. La eliminación por medios mecánicos fue lenta, especialmente en los perros: la semivida fue inicialmente de 140 días y aumentó a 6900 días a partir del día 200 después de la exposición.

En las personas se registró una eliminación inicial rápida del 8% y el 40% de partículas de silicato de aluminio con un diámetro aerodinámico de 1,9 y 6,1 µm, respectivamente, de la región pulmonar durante seis días. A continuación, se eliminaron el 4% y el 11% de los dos tamaños de partículas tras una semivida de 20 días y el resto con semividas de 330 y 420 días.

Las partículas ultrafinas (<100 nm) tienen una deposición alta en la zona nasal y pueden superar la barrera alveolar/capilar.

5. Efectos en mamíferos de laboratorio y en sistemas de ensayo *in vitro*

Un factor determinante importante de la toxicidad de las arcillas es el contenido en cuarzo. La presencia de cuarzo en las arcillas estudiadas dificulta la realización de una estimación independiente fidedigna de la fibrogenicidad de otros componentes.

La inyección intratraqueal única en roedores de bentonita y montmorillonita con un contenido bajo de cuarzo produjo efectos citotóxicos dependientes de la dosis y el tamaño de las partículas, así como inflamación local transitoria, cuyos signos incluían el edema y, en consecuencia, un aumento de peso del pulmón. Las exposiciones intratraqueales únicas de ratas a la bentonita produjeron focos de acumulación en los pulmones 3-12 meses más tarde. Tras la exposición intratraqueal de ratas a bentonita con un contenido elevado de cuarzo también se observó fibrosis. La bentonita aumentó la susceptibilidad de los ratones a la infección pulmonar.

Son limitados los datos sobre los efectos de exposiciones múltiples de animales experimentales a la montmorillonita o la bentonita. Los ratones que recibieron alimentos con un 10% o un 25% de bentonita, pero por lo demás adecuados para un crecimiento normal, presentaron tasas de crecimiento ligeramente reducidas, mientras que en los ratones alimentados de manera semejante, pero con un 50% de bentonita, se registró un crecimiento mínimo y la formación de hígados grasos y en último término de fibrosis del hígado y hepatomas benignos (véase *infra*).

Los estudios *in vitro* de los efectos de la bentonita en diversos tipos de células de mamíferos normalmente pusieron de manifiesto un alto grado de citotoxicidad. Las concentraciones inferiores a 1,0 mg/ml de partículas de bentonita y montmorillonita con un diámetro inferior a 5 µm produjeron daños en las membranas, e incluso lisis celular, así como cambios funcionales en varios tipos de células. La velocidad y el grado de la lisis de los eritrocitos de oveja eran dependientes de la dosis.

El caolín instilado por vía intratraqueal produce focos de acumulación, reacción ante cuerpos extraños y reacción exudativa difusa. En algunos estudios se ha descrito fibrosis tras la administración de dosis elevadas de caolín (con un 8-65% de cuarzo), mientras que, con la administración de dosis más bajas, no se ha observado fibrosis en los pocos estudios disponibles.

La información sobre la toxicidad de la ilita es muy limitada y es nula sobre la de otros componentes de otras arcillas comercialmente importantes. La instilación intratraqueal de ilita con un contenido desconocido de cuarzo indujo proteinosis alveolar, aumento del peso de los pulmones y síntesis de colágeno. La ilita tuvo una citotoxicidad limitada para los macrófagos peritoneales y actividad hemolítica *in vitro*.

No se dispone de estudios adecuados sobre la carcinogenicidad de la bentonita. En un estudio de inhalación y en otro de inyección intrapleural, el caolín no indujo tumores en las ratas. No hay estudios sobre la genotoxicidad de las arcillas.

En estudios únicos muy limitados en ratas no se demostró la toxicidad en el desarrollo tras la exposición oral a la bentonita o el caolín.

6. Efectos en el ser humano

La exposición de la población general a concentraciones bajas de montmorillonita y caolinita, que son los principales componentes de la bentonita y el caolín, respectivamente, y de otros minerales arcillosos es ubicua. No hay información sobre los posibles efectos de dicha exposición a concentraciones bajas.

Las exposiciones ocupacionales prolongadas al polvo de bentonita pueden provocar daños estructurales y funcionales en los pulmones. Sin embargo, los datos disponibles son insuficientes para establecer de manera concluyente una relación dosis-respuesta o incluso relaciones de causa y efecto, debido a la escasa información sobre el tiempo y la intensidad de la exposición y a factores de confusión, tales como la exposición al silicio y al humo del tabaco.

La exposición prolongada al caolín provoca la aparición de neumoconiosis diagnosticada radiológicamente, relacionada con la exposición. Solamente se han notificado un deterioro claramente definido de la función respiratoria y síntomas conexos en casos con resultados radiológicos importantes. La composición de la arcilla, es decir, la cantidad y calidad de los minerales distintos de la caolinita, es un factor determinante importante de los efectos.

La bentonita, el caolín y otras arcillas contienen con frecuencia cuarzo y la exposición al cuarzo tiene una relación causal con la silicosis y el cáncer de pulmón. Se han notificado aumentos estadísticamente significativos de la incidencia de bronquitis crónica y enfisema pulmonar o de mortalidad derivada de estas patologías tras la exposición al cuarzo.

7. Efectos en otros organismos en el laboratorio y en el medio ambiente

La bentonita y el caolín tienen una toxicidad baja para las especies acuáticas, habiéndose sometido a prueba un gran número de ellas.

8. Evaluación del riesgo para la salud humana y de los efectos en el medio ambiente

Según los limitados datos disponibles de estudios sobre personas expuestas a la bentonita, la montmorillonita retenida sólo parece inducir cambios tisulares no específicos ligeros, que son semejantes a los descritos en el espectro de cambios de la "enfermedad de los conductos aéreos pequeños debida al polvo mineral" (acumulaciones nodulares de polvo con material refringente [montmorillonita] en la zona peribronquiolar, junto con fibrosis intersticial limitada). En algunos de los estudios se han notificado asimismo anomalías radiológicas.

No hay casos notificados de reacción fibrótica difusa/nodular acentuada del tejido pulmonar a la montmorillonita en ausencia de silicio libre. No se pueden derivar estimaciones cuantitativas de la capacidad de la bentonita para provocar efectos adversos en los pulmones.

La exposición prolongada al caolín puede dar lugar a una neumoconiosis relativamente benigna, conocida como caolinosis. Se ha observado un deterioro de la función pulmonar sólo en los casos en que había alteraciones radiológicas importantes. Basándose en los datos obtenidos de trabajadores del caolín en el Reino Unido, se puede estimar de manera muy aproximada que su potencia es como mínimo de un orden de magnitud más bajo que la del cuarzo.

La bentonita, el caolín y otras arcillas contienen con frecuencia cuarzo, componente capaz de provocar silicosis y cáncer de pulmón.

No se han encontrado informes sobre los efectos adversos locales o sistémicos debidos al uso frecuente de bentonita o caolín en los cosméticos.

Los efectos biológicos de los minerales arcillosos dependen de su composición mineral y del tamaño de las partículas. El orden decreciente de la potencia del cuarzo, la caolinita y la montmorillonita para producir daño en los pulmones está en consonancia con sus zonas relativas de superficie activa y con la química de la superficie.

La bentonita y el caolín tienen una toxicidad baja para los organismos acuáticos.

THE ENVIRONMENTAL HEALTH CRITERIA SERIES (continued)

Ethylene oxide (No. 55, 1985)
Extremely low frequency (ELF) fields
(No. 36, 1984)
Fenitrothion (No. 133, 1992)
Fenvalerate (No. 95, 1990)
Flame retardants: a general introduction
(No. 192, 1997)
Flame retardants: tris(chloropropyl)
phosphate and tris(2-chloroethyl)
phosphate (No. 209, 1998)
Flame retardants: tris(2-butoxyethyl)
phosphate, tris(2-ethylhexyl) phosphate
and tetrakis(hydroxymethyl)
phosphonium
salts (No. 218, 2000)
Fluorides (No. 227, 2001)
Fluorine and fluorides (No. 36, 1984)
Food additives and contaminants in food,
principles for the safety assessment of
(No. 70, 1987)
Formaldehyde (No. 89, 1989)
Fumonisin B$_1$ (No. 219, 2000)
Genetic effects in human populations,
guidelines for the study of (No. 46, 1985)
Glyphosate (No. 159, 1994)
Guidance values for human
exposure limits (No. 170, 1994)
Heptachlor (No. 38, 1984)
Hexachlorobenzene (No. 195, 1997)
Hexachlorobutadiene (No. 156, 1994)
Alpha- and beta-hexachlorocyclohexanes
(No. 123, 1992)
Hexachlorocyclopentadiene
(No. 120, 1991)
n-Hexane (No. 122, 1991)
Human exposure assessment
(No. 214, 2000)
Hydrazine (No. 68, 1987)
Hydrogen sulfide (No. 19, 1981)
Hydroquinone (No. 157, 1994)
Immunotoxicity associated with exposure
to chemicals, principles and methods for
assessment (No. 180, 1996)
Infancy and early childhood, principles for
evaluating health risks from chemicals
during (No. 59, 1986)
Isobenzan (No. 129, 1991)
Isophorone (No. 174, 1995)
Kelevan (No. 66, 1986)
Lasers and optical radiation (No. 23,
1982)
Lead (No. 3, 1977)[a]
Lead, inorganic (No. 165, 1995)
Lead – environmental aspects
(No. 85, 1989)
Lindane (No. 124, 1991)
Linear alkylbenzene sulfonates
and related
compounds (No. 169, 1996)
Magnetic fields (No. 69, 1987)
Man-made mineral fibres (No. 77, 1988)
Manganese (No. 17, 1981)

Mercury (No. 1, 1976)[a]
Mercury – environmental aspects
(No. 86, 1989)
Mercury, inorganic (No. 118, 1991)
Methanol (No. 196, 1997)
Methomyl (No. 178, 1996)
2-Methoxyethanol, 2-ethoxyethanol, and
their acetates (No. 115, 1990)
Methyl bromide (No. 166, 1995)
Methylene chloride
(No. 32, 1984, 1st edition)
(No. 164, 1996, 2nd edition)
Methyl ethyl ketone (No. 143, 1992)
Methyl isobutyl ketone (No. 117, 1990)
Methylmercury (No. 101, 1990)
Methyl parathion (No. 145, 1992)
Methyl tertiary-butyl ether (No. 206, 1998)
Mirex (No. 44, 1984)
Morpholine (No. 179, 1996)
Mutagenic and carcinogenic chemicals,
guide to short-term tests for detecting
(No. 51, 1985)
Mycotoxins (No. 11, 1979)
Mycotoxins, selected: ochratoxins,
trichothecenes, ergot (No. 105, 1990)
Nephrotoxicity associated with exposure
to chemicals, principles and methods for
the assessment of (No. 119, 1991)
Neurotoxicity associated with exposure to
chemicals, principles and methods for the
assessment of (No. 60, 1986)
Neurotoxicity risk assessment for human
health, principles and approaches
(No. 223, 2001)
Nickel (No. 108, 1991)
Nitrates, nitrites, and N-nitroso
compounds
(No. 5, 1978)[a]
Nitrobenzene
(No. 230, 2003)
Nitrogen oxides
(No. 4, 1977, 1st edition)[a]
(No. 188, 1997, 2nd edition)
2-Nitropropane (No. 138, 1992)
Nitro-and nitro-oxypolycyclic aromatic
hydrocarbons, selected (No.229, 2003)
Noise (No. 12, 1980)[a]
Organophosphorus insecticides:
a general introduction (No. 63, 1986)
Palladium (No. 226, 2001)
Paraquat and diquat (No. 39, 1984)
Pentachlorophenol (No. 71, 1987)
Permethrin (No. 94, 1990)
Pesticide residues in food, principles for
the toxicological assessment of
(No. 104, 1990)
Petroleum products, selected
(No. 20, 1982)
Phenol (No. 161, 1994)
d-Phenothrin (No. 96, 1990)
Phosgene (No. 193, 1997)

---

[a] Out of print

THE ENVIRONMENTAL HEALTH CRITERIA SERIES (continued)

Phosphine and selected metal phosphides (No. 73, 1988)
Photochemical oxidants (No. 7, 1978)
Platinum (No. 125, 1991)
Polybrominated biphenyls (No. 152, 1994)
Polybrominated dibenzo-*p*-dioxins and dibenzofurans (No. 205, 1998)
Polychlorinated biphenyls and terphenyls (No. 2, 1976, 1st edition)[a] (No. 140, 1992, 2nd edition)
Polychlorinated dibenzo-*p*-dioxins and dibenzofurans (No. 88, 1989)
Polycyclic aromatic hydrocarbons, selected non-heterocyclic (No. 202, 1998)
Progeny, principles for evaluating health risks associated with exposure to chemicals during pregnancy (No. 30, 1984)
1-Propanol (No. 102, 1990)
2-Propanol (No. 103, 1990)
Propachlor (No. 147, 1993)
Propylene oxide (No. 56, 1985)
Pyrrolizidine alkaloids (No. 80, 1988)
Quintozene (No. 41, 1984)
Quality management for chemical safety testing (No. 141, 1992)
Radiofrequency and microwaves (No. 16, 1981)
Radionuclides, selected (No. 25, 1983)
Reproduction, principles for evaluating health risks associated with exposure to chemicals (No. 225, 2001)
Resmethrins (No. 92, 1989)
Synthetic organic fibres, selected (No. 151, 1993)
Selenium (No. 58, 1986)
Styrene (No. 26, 1983)
Sulfur oxides and suspended particulate matter (No. 8, 1979)

Tecnazene (No. 42, 1984)
Tetrabromobisphenol A and derivatives (No. 172, 1995)
Tetrachloroethylene (No. 31, 1984)
Tetradifon (No. 67, 1986)
Tetramethrin (No. 98, 1990)
Thallium (No. 182, 1996)
Thiocarbamate pesticides: a general introduction (No. 76, 1988)
Tin and organotin compounds (No. 15, 1980)
Titanium (No. 24, 1982)
Tobacco use and exposure to other agents (No. 211, 1999)
Toluene (No. 52, 1986)
Toluene diisocyanates (No. 75, 1987)
Toxicity of chemicals (Part 1), principles and methods for evaluating the (No. 6, 1978)
Toxicokinetic studies, principles of (No. 57, 1986)
Tributyl phosphate (No. 112, 1991)
Tributyltin compounds (No. 116, 1990)
Trichlorfon (No. 132, 1992)
1,1,1-Trichloroethane (No. 136, 1992)
Trichloroethylene (No. 50, 1985)
Tricresyl phosphate (No. 110, 1990)
Triphenyl phosphate (No. 111, 1991)
Tris- and bis(2,3-dibromopropyl) phosphate (No. 173, 1995)
Ultrasound (No. 22, 1982)
Ultraviolet radiation (No. 14, 1979, 1st edition) (No. 160, 1994, 2nd edition)
Vanadium (No. 81, 1988)
Vinyl chloride (No. 215, 1999)
Vinylidene chloride (No. 100, 1990)
White spirit (No. 187, 1996)
Xylenes (No. 190, 1997)
Zinc (No. 221, 2001)

THE CONCISE INTERNATIONAL CHEMICAL ASSESSMENT SERIES

CICADs are IPCS risk assessment documents that provide concise but critical summaries of the relevant scientific information concerning the potential effects of chemicals upon human health and/or the environment

Acrolein (No. 43, 2002)
Acrylonitrile (No. 39, 2002)
Azodicarbonamide (No. 16, 1999)
Arsine: human health aspects (No. 47, 2002)
Asphalt (bitumen) (No. 59, 1004)
Barium and barium compounds (No.33, 2001)
Benzoic acid and sodium benzoate (No. 26, 2000)

Benzyl butyl phthalate (No. 17, 1999)
Beryllium and beryllium compounds (No. 32, 2001)
Biphenyl (No. 6, 1999)
Bromoethane (No. 42, 2002)
1,3-Butadiene (No. 30, 2001)
2-Butoxyethanol (No. 10, 1998)
Carbon disulfide (No. 46, 2002)
Chloral hydrate (No. 25, 2000)
Chlorinated naphthalenes (No. 34, 2001)

---

[a]   Out of print